Women in Engineering and Science

More information about this series at http://www.springer.com/series/15424

Laura S. Privalle

Editor

Women in Sustainable Agriculture and Food Biotechnology

Key Advances and Perspectives on Emerging Topics

 Springer

Editor
Laura S. Privalle
Bayer CropScience
Morrisville, NC
USA

ISSN 2509-6427 ISSN 2509-6435 (electronic)
Women in Engineering and Science
ISBN 978-3-319-84847-1 ISBN 978-3-319-52201-2 (eBook)
DOI 10.1007/978-3-319-52201-2

Printed on acid-free paper

This Springer imprint is published by Springer Nature
The registered company is Springer International Publishing AG
The registered company address is: Gewerbestrasse 11, 6330 Cham, Switzerland

Contents

Pioneering Women in Sustainable Agriculture and Food Biotechnology

Jill S. Tietjen and Laura S. Privalle

The original plant breeders were women. When humans moved from the hunter/gatherer phase into the cultivation phase, the women kept the seeds from those plants that grew to be the biggest and the strongest for use the following season. These were not the only traits for which the women were looking. Domestication of wild plants required many generations of those plants and fostered traits that included more robust plants, plants with non-shattering seed pods, seeds that did not become dormant, blooms that flowered synchronously across the population, and larger fruits and grains (Flint-Garcia 2015).

As we think of sustainable agriculture and food biotechnology today, many branches of science have been tapped for the advances that we have experienced. Women contributed to each of these sciences as they evolved and led to plant biology and biotechnology. Let's learn about some of those pioneering women through the ages.

Tapputi (Also Tapputi-Belatekallim)—Perfumist (Second Century BC)

Considered the world's first chemist, Tapputi made perfume and is mentioned on a cuneiform tablet from the second millennium BC in Babylonia. Her perfume contained flowers, oil, calamus, cyperus, myrrh and balsam to which she added water. This mixture was then distilled and filtered in her still; the oldest referenced

J.S. Tietjen
Greenwood Village, Colorado, USA
e-mail: jill.s.tietjen@gmail.com

L.S. Privalle (✉)
Research Triangle Park, North Carolina, USA
e-mail: laura.privalle@bayer.com

© Springer International Publishing AG 2017
L.S. Privalle (ed.), *Women in Sustainable Agriculture and Food Biotechnology*,
Women in Engineering and Science, DOI 10.1007/978-3-319-52201-2_1

still of which we are aware. Women perfumers used the chemical techniques of distillation, extraction and sublimation to create their perfumes, which were important in medicines and religion as well as for cosmetics (Alic 1986).

Miriam the Alchemist (1st or 2nd Century AD)

Born in Alexandria, Egypt, Miriam was also known as Mary, Maria, and Miriam the Prophetess or the Jewess. Her major inventions and improvements included the three-armed still or *tribikos*, the *kerotakis*, and the water bath. Although the purpose of the inventions was to accelerate the process of metals transmuting into gold, they are used extensively in modern science and contemporary households. The *tribikos* was an apparatus for distillation, a process of heating and cooling that imitated processes in nature. Sponges formed a part of the mechanism and served as coolers. The *kerotakis* was an apparatus named for the triangular palette used by artists to keep their mixtures of wax and pigment hot. The water bath, also known as Marie's bath (*bain-marie*), is similar to the present-day double boiler (B.F. Shearer and B.S Shearer 1997).

Hildegard of Bingen—Natural Philosopher (1098–1179)

A Benedictine abbess known as "the Sibyl of the Rhine," Hildegard wrote music as well as treatises on science including cosmology, medicine, botany, zoology, and geology. Two of her manuscripts, *Causae et curae* (*Causes and Cures* or *Book of Compound Medicine*) and *Physica* (*Natural History* or *Book of Simple Medicine*) are considered among the greatest scientific works of the Middle Ages and have survived intact. *Physica* is her natural history textbook and included descriptions of nearly 500 plants, metals, stones and animals, and explains their medicinal value to humans. The book became a medical school text. In *Causae et curae*, Hildegard describes the relationships between the macrocosm and specific diseases of the microcosm, the human body, and prescribed medicinal remedies. Hildegard was the first medical writer to stress the importance of boiling drinking water (Ambrose et al. 1997; Proffitt 1999; B.F. Shearer and B.S Shearer 1996).

Marie Meurdrac—Alchemist (C. 1610–1680)

Marie Meurdrac was not aware of Miriam the Alchemist's chemistry work when she wrote a six-part chemistry treatise. Meurdrac covered laboratory principles, apparatus and techniques, animals, metals, the properties and preparation of medicinal simple and compound medicines, and cosmetics. Her work included a table of weights as well as 106 alchemical symbols. Her work titled *La Chymie charitable et facile en faveur des dames* was first published in Paris in 1666. Later editions were issued in 1680 and 1711. Her foreword to her book contained the following thought: … *that minds have no sex and that if the minds of women were cultivated like those of men, and if as much time and energy were used to instruct the minds of the former, they would equal those of the latter* (Alic 1986).

Jane Colden—Botanist (1724–1766)

By 1757, Jane Colden (later Farquhar), the first woman botanist in the U.S., had prepared a catalog of over 300 local species of flora and had exchanged specimens and seeds with several colonial and European botanists. Under the tutelage of her father, Cadwallader Colden, a New York botanist and government leader, Jane Colden mastered the Linnaean classification system and wrote a paper for a publication by the Edinburgh Philosophical Society. She is best known for her identification and description of the gardenia, which she was the first to identify. Her botanical work ceased after her marriage in 1759 (Ambrose et al. 1997; Bailey 1994; Rossiter 1992; Ogilvie 1993).

Marie Anne Pierrette Paulze Lavoisier—Chemist (1758–1836)

Antoine and Marie Lavoisier established chemistry as a modern scientific discipline. Their discoveries included the identification of oxygen and the nature of combustion, oxidation, and respiration. In addition, they established the law of conservation of matter as a principle for experimental design. It is impossible to separate Marie's contributions from Antoine's although she is known to have assisted with experiments and kept all of the laboratory records and notes. She edited and illustrated her husband's treatise *Elements of Chemistry* (1789) and translated and wrote commentaries on scientific papers, including Richards Kirwan's 1787 *Essay on Phlogiston* (Proffitt 1999; Ogilvie 1993).

Jane Haldimand Marcet—Science Popularizer (1769–1858)

Remembered particularly for the impact her *Conversations in Chemistry* had on influencing future scientist Michael Faraday, Jane Marcet wrote books to popularize science, especially intended for women and young people. Marcet was encouraged to begin a writing career by her husband, physician Dr. Alexander Marcet, whose passion for chemistry exceeded his interest in practicing as a physician. *Conversations in Chemistry* (1806) was very popular and went through numerous editions, including 15 American editions titled *Mrs. Bryan's Conversations*. Marcet believed that the information presented in a conversational format was more readily comprehended by the audience, as she was better able to understand chemistry after conversing with a friend. Her other books included *Conversations on Botany, Conversations on Natural Philosophy, Conversations on Political Economy*, and *Conversations on Vegetable Physiology* (Proffitt 1999; Ogilvie 1993; Ronan 1982; Suplee 2000; A dictionary of scientists 1999).

Josephine Ettel Kablick (Josefina Kablíková)—Botanist and Paleontologist (1787–1863)

An intrepid Czech botanist and paleontologist, Josephine Kablick collected plant and fossil samples. Undeterred by any weather or terrain, she gathered new species in dark forests and on mountains. Her collection gained renown and she gradually collected plants for schools and colleges in her country as well as for museums and learned societies in other parts of Europe. Fittingly, many of the fossils and plants that she collected are named in her honor (Mozans 1913).

Estella Eleanor Carothers—Zoologist and Cytologist (1882–1957)

Eleanor Carothers studied cells and their inner workings. She particularly examined the relationship between cytology and genetics with specific emphasis on the effects that X-rays have on cells. Through her research thoroughness and her emphasis on the genetics of the order Orthoptera (including crickets and grasshoppers), she answered many questions concerning cytological heredity. Considered a primary investigator in the field of genetics, Carothers focused on grasshopper embryos. Her name is starred in the 1926 edition of the *American Men of Science*, meaning that she was considered one of the foremost scientists of the day. Carothers received many honors including the 1921 Ellen Richards Research Prize from the Naples Table Association and election to the National Academy of Sciences (Proffitt 1999; B.F. Shearer and B.S Shearer 1996; Ogilvie 1993).

Gerty Cori—Biochemist (1896–1957)

Nobel Laureate Gerty Cori was the first American woman to win a Nobel Prize in science. She and her husband, Dr. Carl Ferdinand Cori, received the 1947 Nobel Prize in Physiology or Medicine "for their discovery of the course of the catalytic conversion of glycogen." They explained the physiological process by which the body metabolizes sugar.

Cori was born in Prague where her uncle, a professor in pediatrics, nurtured her interest in mathematics and science and encouraged her to undertake the study necessary to enter a university and study medicine. By age 18, she had passed a very difficult examination and entered the German branch of the medical school at Prague's Carl Ferdinand University. During her first semester anatomy class, she met her husband-to-be. They jointly agreed to pursue medical research, not medical practice, and to jointly attain medical certification (a 6-year process) before marrying.

In 1922, Carl received an offer to work in the U.S. and Gerty, demonstrating significant independence, stayed behind until she too had an offer to work in the U.S. They both worked at the New York State Institute for the Study of Malignant Diseases (later the Roswell Park Memorial Institute) in Buffalo, New York; he as a biochemist, she as an assistant pathologist. Here, Cori experienced resistance to her presence as a woman in science. The director of the institute threatened to fire Gerty if she did not end collaborative work with her husband. Later, a university offered Carl a job—only if he ended working collaboratively with his wife. The rationale for these requests was that not only was it un-American for a man to work with his wife—his wife was standing in the way of his career advancement!

Not everyone believed this however. After becoming naturalized American citizens in 1928, Gerty and Carl received offers to work at Washington University in St. Louis. Carl would become a professor of pharmacology and Gerty was offered the position of research associate in pharmacology. Here, Gerty gave birth to their son, Thomas who eventually became a research chemist himself, following in his parents' footsteps.

Although denied positions and titles that she would have received as a man, Gerty was promoted to associate professor in biochemistry in 1943, the year she and Carl achieved the synthesis of glycogen in a test tube. In 1947, shortly before she was awarded the Nobel Prize, Gerty was promoted to full professor of biochemistry. The Cori's discovery of glycogen led to more effective treatments for diabetes. The relationships between the liver and muscle glycogen, and blood glucose and lactic acid is now known as the Cori cycle. Gerty's other areas of research included hereditary glycogen storage diseases in children and the identification of a new enzyme, amylo-1, 6-glucosidase which helped her identify the structure of glycogen. She became a member of the National Academy of Sciences in 1948 (B.F. Shearer and B.S. Shearer 1997; Proffitt 1999; Bailey 1994; Kass-Simon and Farnes 1990; McGrayne 1993).

Barbara McClintock—Geneticist (1902–1992)

Barbara McClintock received the Nobel Prize in Physiology or Medicine in 1983 for her discovery that genes can move around on the chromosomes (transposable elements)—the so-called "jumping genes." She first published the discovery in

1950, but it was not accepted in the scientific community for many years and she worked on her research for many years alone. Her novel idea took 35 years for the Nobel Prize because it was such a revolutionary concept. In addition, the transposable elements that she had conjectured weren't actually seen until the late 1970s when the science of molecular biology had developed significantly further than it had as of 1950.

McClintock was recognized as one of the brightest geneticists from her graduate school days at Cornell in the 1920s. After serving as an instructor in botany for 5 years and then working in research for 6 years, she left Cornell as they would not appoint women to faculty positions. In the early 1930s, she found chromosomes that formed rings. Later, she found that the ring chromosomes were a special case of broken chromosomes. She predicted the existence of structures, which she named telomeres, that would be found on the ends of normal chromosomes, that maintained a chromosome's stability and integrity but were lost when a chromosome was broken. Telomere research is a rapidly growing area of biology today, with specific implications for cancer and aging. McClintock served as an assistant professor of botany at the University of Missouri for 5 years. In 1942, she began work at the Cold Spring Harbor Laboratory on Long Island, New York where she would spend the rest of her career.

McClintock was recognized for her genetic work, however, even if the Nobel Prize was slow in coming (the general span is 10–15 years after the research or discovery). Her name is starred (indicating eminence as a scientist) in the seventh edition of *American Men of Science*. She was elected the first woman president of the Genetics Society of America in 1945. In 1944, she was elected to the National Academy of Sciences. McClintock received the National Medal of Science in 1970. She also received awards including the Kimber Genetics Award (1967), the Lasker Award (1981), and a MacArthur genius award starting in 1981 (Proffitt 1999; B.F. Shearer and B.S Shearer1996; Bailey 1994; McGrayne 1993).

Rosalind Franklin—Biologist (1920–1958)

Rosalind Franklin made key contributions to the structures of coals and viruses and provided the scientific evidence about the double-helix structure of DNA for which James Watson, Francis Crick, and Maurice Wilkins shared the Nobel Prize in 1962. Although Nobel Prizes are only awarded to living scientists, her contributions to the effort to discover the structure of DNA are thought by some to have been overlooked.

Franklin grew up in London and decided at any early age to pursue a career in science. She graduated from Cambridge in 1941 and after a short-lived research scholarship to study gas-phase chromatography with future Nobel laureate Ronald G.W. Norrish, accepted a job as assistant research officer with the British Coal Utilization Research Association (CURA). At the CURA, she applied her knowledge of physical chemistry to study the microstructures of coal. In 1947, she moved to Paris where she learned the technique known as X-ray diffraction. In 1951, she left Paris to set up an X-ray diffraction unit in a laboratory at St. John T. Randall's Medical Research Council at Kings' College in London to produce diffraction pictures of DNA.

Here she worked with Maurice Wilkins, who took an intense dislike to her. Wilkins would later show Watson the DNA diffraction pictures that Franklin had amassed (without her permission) and here Watson saw the evidence needed to discern the helical structure of DNA. Franklin had recorded in her laboratory notebook that DNA had a helical structure of two chains prior to the publication by Watson and Crick of their similar analysis.

Franklin left King's College for Birkbeck College where she worked on the tobacco mosaic virus, particularly the RNA structure and the location of protein units. She died at age 37 from ovarian cancer (Proffitt 1999; B.F. Shearer and B.S Shearer 1996; McGrayne 1993).

Indra and Vimla Vasil—Plant Biotechnologists

After obtaining their Ph.Ds. from the University of Delhi in 1958 and 1959, respectively, Indra and Vimla Vasil came to the U.S. on sabbaticals in the early 1960s and worked with A.C. Hildebrant. There, Vimla demonstrated the totipotency of plants cells by regenerating plants from single cells of tobacco. Her husband, Indra, demonstrated that plant species other than carrot could form somatic embryos. These two pioneers were both at the University of Florida from 1967 to 1999 where they worked together on in vitro biology and biotechnology of cereals. Their production of the first detailed account of embryonic cultures of cereals led to successful regeneration in numerous cereals and grasses. The Vasils were the first to obtain transgenic wheat using biolistic technology. In 2007, they jointly received the Society for In Vitro Biology's Lifetime Achievement Award. In their acceptance remarks, the Vasils said, "Based on our own experiences we feel that it is important for senior scientists to provide support and guidance to the next generation of students, and to encourage them to think big, think bold, think different, and not be afraid to challenge conventional wisdom and dogmas" (https://sivb.org/InVitroReport/41-3/lifetime.htm).

Norma Trolinder—Geneticist

A pioneering cotton research geneticist, Norma Trolinder, Ph.D. founded Genes Plus, a research company specializing in genetic engineering work. She was also president and research director of Southplains Biotechnology, Inc. and a research scientist for 8 years at the USDA Cropping Systems Research Lab in Lubbock, Texas. Her work together with her daughter Linda Trolinder (today Head of Trait

Development—Cotton, Corn and Soy for Bayer CropScience) on cotton transformation and regeneration was critical to being able to successfully produce commercial transgenic cotton such at Bt^1 cotton. Upon her receipt of the 2000 Cotton Genetics Research Award, it was said "her diligent efforts in the difficult area of plant regeneration from cotton tissue overcame a major hurdle in cotton biotechnology. Her work was essential to the successful utilization of transgenic cotton in the industry that we are experiencing today." Trolinder's bachelor's, master's and doctorate degrees are from Texas Tech University (http://www.cotton.org/news/releases/2001/cotton-genetics-research-award.cfm).

Mary-Dell Chilton-Plant Biotechnologist (1939–)

In 1983, Mary-Dell Chilton led the research team that produced the first transgenic plants. As such, she is considered one of the founders of modern plant biotechnology and the field of genetic engineering in agriculture. After groundbreaking efforts at the University of Washington and Washington University, she established one of the world's leading industrial biotechnology agricultural programs at Ciba-Geigy (today Syngenta). Her team has worked to produce crops with higher yields, and resistance to pests, disease and adverse environmental conditions (such as drought).

The recipient of numerous awards including the 1985 Rank Prize in Nutrition and the 2013 World Food Prize, Chilton was inducted into the National Inventors Hall of Fame in 2015. Today, Distinguished Science Fellow Chilton works in a building in the Research Triangle Park in North Carolina that bears her name.

Dr. Chilton's B.S. and Ph.D. degrees are in chemistry from the University of Illinois Urbana-Champaign. She said "My career in biotechnology has been an exciting journey and I am amazed to see the progress we have made over the years. My hope is through discoveries like mine and the discoveries to follow, we will be able to provide a brighter and better future for the generations that follow us" (Lacapra 2015; http://invent.org/inductees/chilton-mary-dell/; http://www.worldfoodprize.org/index.cfm/24667/35489/syngenta_scientist_dr_marydell_chilton_named_2015_national_inventors_hall_of_fame_inductee).

Barbara Hohn—Molecular Biologist (1939–)

Barbara Hohn considers herself privileged to have witnessed and contributed to major steps in the understanding of the molecular basis of life. She was involved in cloning of DNA and transformation of and genetic recombination in plants.

---

[1]Bt as a modifier before a plant means that the plant has been genetically altered to express proteins from the bacterium *Bacillus thuringiensis*.

Reveling in naiveté and curiosity, her research into basic principles led to the discovery that maize (corn) could be the host for DNA transfer. In addition to working at the FMI Institute for Biomedical Research in Basel Switzerland, Hohn's career included time at Yale, Stanford, and the University of Basel. The recipient of many awards, Hohn studied chemistry and received her Ph.D. in biochemistry (http://www.fmi.ch/about/people/emeriti/emeriti.html?group=14).

Martha Wright—Biologist

Although she entered the Kansas City Science Fair in 1956, Martha Wright's father encouraged her to major in business at then Lindenwood College in St. Charles, Missouri, because she would always be able to get a job as a secretary. A business advisor, noting that Wright was bored, urged her to sign up for an advanced biology course—and she was hooked. Graduating with a biology major (and minors in chemistry, classics, and business), Wright was hired by Monsanto because she had worked with radioactivity while in college (one of her biology professors had worked on the Manhattan Project). Her early projects revolved about insecticides. Then, she became involved in the pioneering work on field crop cell culture, working particularly with soybeans, maize and alfalfa. She published papers on regenerating soybeans from cell culture. After joining what is today Syngenta, her attention turned to corn and pioneering work on that cell culture. Her team produced the event that became the first commercial Bt corn product, Bt 176 (also known as Maximizer Knockout™). Wright says "our work broke the mystique of plant regeneration from cell culture, and ultimately allowed the transformation of recalcitrant crops. Enhanced crops mean more people get to eat and more people are healthy and can devote their energies to improving the world" (Neal Stewart 2008).

Nina Fedoroff—Molecular Biologist (1942–)

The recipient of the National Medal of Science for "pioneering work on plant molecular biology and for her being the first to clone and characterize maize transposons," Nina Fedoroff is Emeritus Professor of Biology at Penn State University. Her research interests include plant stress response, hormone signaling, transposable elements, and epigenetic mechanisms. An expert in the fields of plant genetics and molecular biology, she joined the faculty at UCLA after receiving her Ph.D. in molecular biology where she did research on nuclear RNA. As one of the first plant molecular biologists, Fedoroff pioneered DNA sequencing while working at the Carnegie Institution for Science. Later, she worked on the molecular characterization of jumping genes (transposable elements—for which Barbara McClintock won the Nobel Prize). A member of the National Academy of Sciences, Fedoroff has served on the National Science Board and received many honors (http://bio.psu.edu/news-and-events/

2008; http://bio.psu.edu/directory/nvf1; https://en.wikipedia.org/wiki/Nina_Federoff; http://www.ofwlaw.com/attorneys/dr-nina-v-fedoroff/).

Virginia Walbot—Agriculturist and Botanist (1946–)

Virginia Walbot loved striped flowers at an early age, so it seems not surprising that her career has focused on the characteristics of the striped and speckled seeds of Indian corn. After undergraduate work at Stanford and graduate work at Yale, she spent time as a faculty member at Washington University in St. Louis. There, she began her work with maize, spending time with Nobel Laureate Barbara McClintock at Cold Spring Harbor. Today, Walbot is a Professor of Biology at Stanford University. She is a member of the team that developed a new sweet corn. A Fellow of the American Association for the Advancement of Science, Walbot has received many awards and honors and was the first foreign woman elected as a corresponding member of the Mexican Academy of Sciences (in 2004). Walbot is concerned about scientific literacy and presents many lectures at which she encourages a discussion of the science underlying transgenic food (http://www.k-state.edu/bmb/seminars/hageman/2001-walbot.html; http://web.stanford.edu/∼walbot/cv/cv_walbot.pdf; https://profiles.stanford.edu/virginia-walbot).

Lydia Villa-Komaroff—Molecular Biologist (1947–)

The third Mexican-American woman to earn a Ph.D., Villa-Komaroff was part of the research team that discovered insulin could be produced from bacteria. Inspired by her mother's love of nature and plants and her inability to study botany after a bout of rheumatic fever as a child, Villa-Komaroff studied biology at the University of Washington. After graduating with a Ph.D. from MIT, she focused on the synthesis of eggshell proteins using the new technology of recombinant DNA (combining the DNA from one organism to the DNA of bacteria). She used that technique as a member of the team at Harvard that successfully produced insulin from bacteria. This patented process led to almost all commercial insulin today being made from bacteria. Among her many awards, her favorite is "100 Most Influential Hispanics" (Proffitt 1999).

Patricia Zambryski—Plant Biologist

Professor of plant and microbial biology at the University of California, Berkeley, Patricia Zambryski is a pioneer in the development of genetic engineering in plants. Zambryski discovered how the bacterium *Agrobacterium tumefaciens* transfers

DNA into the plant that it infects. This discovery and additional investigation into this specific bacterium have led to fundamental insights applicable to numerous areas of bacterial and plant biology. Zambryski grew up in Canada, receiving her B.S. in genetics and later received her Ph.D. in molecular biology. A Fellow of the American Association for the Advancement of Science and a Fellow of the American Society for Microbiology, Zambryski was elected to the National Academy of Sciences in 2001. Her current research focuses on studying the molecular mechanisms of *Agrobacterium* that leads to genetic transformation of plant cells. She and her lab also study how plant cells communicate with one another (http://pmb.berkeley.edu/profile/pzambryski; http://www.usias.fr/en/evenements/visitors/martin-sarter/pat-sambryski/).

Barbara Mazur—Agricultural Biotechnologist (1949–)

Currently the Vice President, Technology Acquisition Strategy for DuPont Pioneer, Barbara Mazur has degrees in microbiology and molecular genetics. A long-time DuPont employee, she began her DuPont career in the Central Research Department. Her primary focus has been on the modification of seed quality traits, crop protection biochemical discovery, and herbicide resistance trait discovery and development. Mazur holds four patents and has served on advisory boards for the National Academy of Sciences and the National Science Foundation. She has been featured as a STEM (science, technology, engineering and mathematics) Women All-Star, where her background is described as "A research leader at DuPont working to increase food production by improving the rate of crop seed genetic gain through biotechnology and advanced breeding technologies" (http://www.kgi.edu/about-kgi/board-of-trustees/barbara-j-mazur; http://www2.dupont.com/Media_Center/en_US/assets/downloads/pdf/DuPont_Speakers_Bureau.pdf; http://the-dupont-challenge.tumblr.com/post/41289108646/our-second-dupont-stem-women-all-star-is-barbara).

Anne Knupp Crossway—Biologist, Geneticist, Business Manager (1953–)

The pioneering developer of a laboratory method called micromanipulation, a technique that is now widely used in transferring genetic materials from one cell to another, Anne Knupp Crossway did her undergraduate work in biology. Her education included a Ph.D. in genetics and an MBA. Crossway holds two patents, one for microassay for detection of DNA and Ribonucleic acid and one for plant cell microinjection technique. Early in her career, she worked in venture capital as well as holding management positions in biotechnology, consumer products, and over-the-counter drug/device companies. Crossway also served as a managing scientist at an early successful biotechnology company, Calgene, Inc. As a biotechnology consultant she provides services to clients including bioscience companies in early stages of development and non-profit science and technology companies (https://www2.cortland.edu/bulletin/issues/bulletin_05_06/April_17_06.pdf; http://prabook.com/web/person-view.html?profiledId=792869; http://readme.readmedia.com/SUNY-Cortland-To-Hold-12th-Annual-Scholars-Day-April-9/118007).

Elizabeth Hood—Biologist

Currently Distinguished Professor of Agriculture at Arkansas State University, Elizabeth Hood's primary areas of focus are biomass to biobased products (renewable resources), foreign gene expression in transgenic plants, plant cell wall structure and function, and plant cell biology and protein targeting. With three partners, Hood started a company whose purpose is to produce enzymes for biomass conversion from transgenic plants. Her research laboratory at ASU examines plant-based protein production technology and cell wall structure and function. She has previously served as the Program Director at the National Science Foundation and formed a research group at ProdiGene, a plant biotechnology company. At Pioneer Hi-Bred International, Hood served as the director of the cell biology group for plant production of therapeutic proteins. She has many publications, 14 patents and has received numerous honors. Hood has an M.S. in botany from Oklahoma State University and her Ph.D. is in plant biology from Washington University in St. Louis, Missouri (http://biobasedsolutions.org/#!about-us-page).

Ann Depicker—Plant Biologist

The Group Leader of the Vlaams Institute voor Biotechnologies (VIB) at the University of Ghent (Belgium) since 1996, and Division Coordinator since 2003, Ann Depicker leads the group focused on plant-made antibodies and immunogens. A significant breakthrough from her group's work is the simplification of the process for producing biotech medicines—using plant seeds instead of mammalian cells. This effort was undertaken to create an alternative and cost effective system for the production of complex recombinant proteins including antibodies and composite vaccines. Plant seeds can be produced in large quantities in short periods of time as compared to mammalian cells without the need for specialized equipment or expensive media. The technique is called the GlycoDelete technology. Depicker said "the simplicity of the modification makes an industrial approach possible and could lead to the inexpensive large-scale 'pharming' of medicines using plants" (http://www.vib.be/en/research/scientists/Pages/Ann-Depicker-Lab.aspx; http://www.seedquest.com/solutions.php?type=solution&id_article=76572).

Maud Hinchee—Botanist

Maud Hinchee's botany roots go back to her childhood when she made the eggplants in her mother's backyard garden sterile because she hated eggplant. After receiving BS, MS and Ph.D. degrees in botany, Hinchee became involved with protocols for inserting genes into plant cells due to her training as a plant morphogeneticist. At Monsanto, she was able to design methods that allowed successful and reproducible results in soybean plants—and led to the first transgenic soybean containing the Roundup® Ready gene. Hinchee worked on other plants including sugar beet, flax, potato, strawberry, cotton and sweet potato. After time at ArborGen as Chief Technology Officer where she led efforts to genetically improve tree products, Hinchee moved to Agricen Sciences. As Chief Science Officer at Agricen Sciences, Hinchee is striving to develop solutions for increased nutrient availability and uptake, improved sustainability of agriculture, and higher crop yields. The holder of five patents, Hinchee was named a Monsanto Science Fellow in recognition of her scientific achievements (Neal Stewart 2008; http://www. agricensciences.com/about-us/our-team/maud-hinchee/; http://www.genomecanada.ca/en/about/governance/hinchee.aspx).

Pamela Ronald—Plant Pathologist and Geneticist (1961–)

An international advocate for genetic engineering of food crops, Dr. Pam Ronald is a professor at the University of California, Davis as well as the Director of the Institute for Food and Agricultural Literacy located there and the Director of Grass Genetics at the Joint BioEnergy Institute in Emeryville, California. She determined her career focus, rice (the biggest food staple in the world), during her years as a graduate student at the University of California, Berkeley (from which she received her Ph.D.). Her laboratory has developed rice that is disease-resistant and flood-tolerant. Featured in a TED talk in 2015, her topic was how genetic engineering can fight disease, reduce insecticide use, and enhance food security. Her undergraduate degree is from Reed College (Oregon) which afforded her the opportunity to study the recolonization of Mt. St. Helens. She has a master's degree from Stanford as well as from Uppsala University, Sweden (http://biosci3.ucdavis.edu/Faculty/Profile/View/14069; http://en.wikipedia.og/wiki/Pamela_Ronald; https://www.jbei.org/people/directors/pam-ronald/).

Sustainable Agriculture

Sustainable agriculture, of course, relies on more options than just those offered through the application of biotechnology solutions, however in this volume we have chosen to focus on the contributions that biotechnology products bring to the

Critical Enabling Technologies

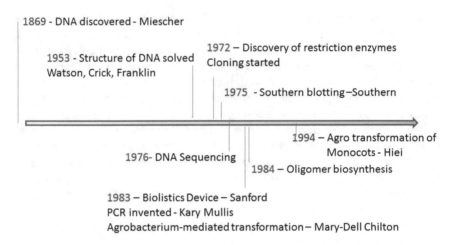

Fig. 1 Key Enabling Technologies. Presented is a time-line with key enabling technologies that were essential to the development of agricultural biotechnology products.

table. These are critical solutions when considering the necessity for sustainable agriculture and the expected population growth the world will see in the next 30 years and beyond. To set the stage, key scientific accomplishments were necessary. A timeline of the critical enabling technologies, as developed by Dr. Privalle, is shown in Fig. 1. All these technologies were necessary to lead to the key early regulatory approvals shown in Table 1, and those of today—but 1983 really stands out. Three food scares occurred in Europe in the 1990s which were key in increasing public mistrust of government agencies as they were used by anti-biotechnology factions to fuel the flames of uncertainty around these technologies. In 1996, the first case of mad cow disease in Europe was reported. In 1999, dioxin contamination was found in animal feed in Belgium and, in 1999, Coca-Cola products were withdrawn in Belgium. These scares resulted in a moratorium (1998–2004) on approvals of biotech products while a new agency was formed for evaluation of not only biotechnology products but also of new products that were entering the market or were associated with food production. To date, only four biotechnology products have been approved for cultivation in Europe: Bt 176, Mon810 and T25, prior to the moratorium; since the moratorium only the Amflora potato has been approved for cultivation. Multiple products have been approved for import into the EU.

Today, biotechnology products are grown in 28 countries around the world with more than 90% of the growers small land holders. (James 2015). Recent approvals show that not only major biotechnology corporations can bring products to market but smaller organizations like EMBRAPA (soybean), Simplot (potato), Okranaga (apple), Mahyco (Bt eggplant) and the USDA (virus resistant plum) are also having

Table 1 Key early US regulatory approvals of biotechnology products (http://cera-gmc.org)

Year	Product	Agency	Registrant
1992	FlavrSavr Tomato (delayed ripening)	FDA	Calgene (now Monsanto)
1994	Bt176 Maize (insect resistant)	EPA (1995, USDA, FDA)	Ciba-Geigy (now Syngenta)
	Bromoxynil Cotton (herbicide tolerant)	USDA, FDA	Calgene (now Monsanto)
	RR Soybean (herbicide tolerant)	USDA (FDA, 1995)	Monsanto
1995	NewLeaf Potato (insect resistant)	USDA, EPA, FDA	Monsanto
	T25 Maize (herbicide tolerant)	USDA	Agrevo (now Bayer)
	RR Canola (herbicide tolerant)	USDA, FDA	Monsanto
	Mon810 (Insect resistant)	EPA (USDA, FDA, 1996)	Monsanto
1996	Virus resistant Papaya	USDA (EPA, FDA 1997)	Cornell University
	InVigor Canola	FDA, USDA	Aventis (now Bayer)
1998	RR sugarbeet	FDA, USDA	Monsanto/Novartis (now Syngenta)

success. The early generation products were virus resistant, herbicide tolerant, and/or insect resistant but now products for drought tolerance, low acrylamide, reduced browning, and golden rice[2] are being developed. Finally products are reaching those areas that most need them like Cameroon, Nigeria, and Kenya. Still there is doubt about the utility, benefits and safety of these products. All the authors in this volume are women who have contributed to or benefited from this technology and have watched its adoption, rejection and debate. Enjoy their stories!

References

Alic M (1986) Hypatia's heritage: a history of women in science from antiquity through the nineteenth century. Beacon Press, Boston

Ambrose SA, Dunkle KL, Lazarus BB, Nair I, Harkus DA (1997) Journeys of women in science and engineering: no universal constants. Temple University Press, Philadelphia

Ann Depicker Lab. http://www.vib.be/en/research/scientists/Pages/Ann-Depicker-Lab.aspx

---

[2]Golden rice is a genetically modified rice that is golden in color due to the over production of β-carotene, the main precursor of Vitamin A. Consumption of golden rice would greatly reduce Vitamin A deficiency and hence reduce blindness and death in children around the world whose diet currently is insufficient in Vitamin A.

Bailey Martha J (1994) American women in science: a biographical dictionary. ABC-CLIO, Denver

Barbara J. Mazur, Ph.D., Vice President, Research Strategy, Agricultural Biotechnology. http://www2.dupont.com/Media_Center/en_US/assets/downloads/pdf/DuPont_Speakers_Bureau.pdf

Center for Environmental Risk Assessment (CERA). http://cera-gmc.org

Dr. Nina Fedoroff receives National Medal of Science. http://bio.psu.edu/news-and-events/2008-news/nina-fedoroff-receives-national-medal-of-science

Dr. Nina V. Fedoroff, Olsson Frank Weeda Terman Matz PC. http://www.ofwlaw.com/attorneys/dr-nina-v-fedoroff/

Elizabeth Hood, Ph.D. http://biobasedsolutions.org/#!about-us-page

Flint-Garcia S (2015) ILSI IFBic—safety of GM crops: compositional analysis. https://www.youtube.com/watch?v=v7P8oVos2jA. Accessed 4 Sept 2015

Hageman Lecturer, Kansas State University. http://www.k-state.edu/bmb/seminars/hageman/2001-walbot.html. April 30–May 1, 2001

Hohn B, Friedrich Miescher Institute for Biomedical Research. http://www.fmi.ch/about/people/emeriti/emeriti.html?group=14

James C (2015) 20th Anniversary (1996 to 2015) of the global commercialization of biotech crops and biotech crop highlights in 2015. ISAAA Brief No. 51. ISAAA, Ithaca

Kass-Simon G, Farnes P (eds) (1990) Women of science: righting the record. Indiana University Press, Bloomington

KGI—Keck Graduate Institute, Barbara J. Mazur, Vice President, Technology Acquisition Strategy, DuPont Pioneer. http://www.kgi.edu/about-kgi/board-of-trustees/barbara-j-mazur

Lacapra V (2015) Interview: Mary-Dell Chilton on her pioneering work on GMO crops, Genetic Literary Project, May 27, 2015, St. Louis Public Radio. http://www.geneticliteracyproject.org/2015/05/27/interview-mary-dell-chilton-on-her-pioneering-work-on-gmo-crops/. Accessed 6 June 2015

Maud Hinchee, Phd, Chief Science Officer, Agricen Sciences. http://www.agricensciences.com/about-us/our-team/maud-hinchee/

Maud A. Hinchee, Chief Technology Officer, ArbonGen, LLC, Genome Canada. http://www.genomecanada.ca/en/about/governance/hinchee.aspx

McGrayne SB (1993) Nobel Prize women in science: their lives, struggles, and momentous discoveries. Carol Publishing Group, New York

Mozans HJ (1913) Woman in science with an introductory chapter on woman's long struggle for things of the mind. D. Appleton and Company (Converted to a kindle edition by John Augustine Kahm)

National Inventors Hall of Fame, Inductees: Mary-Dell Chilton. http://invent.org/inductees/chilton-mary-dell/. Accessed 6 June 2015

Neal Stewart C Jr (ed) (2008) Plant biotechnology and genetics: principles, techniques and applications. John Wiley & Sons

Nina Fedoroff. https://en.wikipedia.org/wiki/Nina_Federoff

Nina V. Fedoroff—Penn State University Department of Biology. http://bio.psu.edu/directory/nvf1

Ogilvie MB (1993) Women in science: antiquity through the nineteenth century, a biographical dictionary with annotated bibliography. MIT Press, Cambridge, pp 60–61

Pamela Ronald. http://biosci3.ucdavis.edu/Faculty/Profile/View/14069

Pamela Ronald. http://en.wikipedia.og/wiki/Pamela_Ronald

Pat Zambryski, USIAS—University of Strasbourg. http://www.usias.fr/en/evenements/visitors/martin-sarter/pat-sambryski/

Patricia C. Zambryski, University of California, Berkeley, College of Natural Resources, Plant and Microbial Biology. http://pmb.berkeley.edu/profile/pzambryski

Prabook, Anne Crossway (born March 22, 1953), American biologist, geneticist, business manager. http://prabook.com/web/person-view.html?profiledId=792869

Proffitt P (ed) (1999) Notable women scientists. The Gale Group, Detroit

Ronan CA (1982) Science: its history and development among the world's cultures. The Hamlyn Publishing Group Limited, New York

Rossiter MW (1992) Women scientists in America: struggles and strategies to 1940. The Johns Hopkins University Press, Baltimore

Shearer BF, Shearer BS (eds) (1996) Notable women in the life sciences: a biographical dictionary. Greenwood Press, Westport

Shearer BF, Shearer, BS (eds) (1997) Notable women in the physical sciences: a biographical dictionary. Greenwood Press, Westport

SIVB Lifetime Achievement Award, In Vitro Report, An Official Publication of the Society for In Vitro Biology, Issue 41.3, July–September 2007. https://sivb.org/InVitroReport/41-3/lifetime.htm

SUNY Cortland to Hold 12th Annual Scholar's Day April 9. http://readme.readmedia.com/SUNY-Cortland-To-Hold-12th-Annual-Scholars-Day-April-9/118007

Suplee C (2000) Milestones of science. National Geographic, Washington

Texas Research Geneticist Receives 2000 Cotton Genetics Research Award, National Cotton Council of America. http://www.cotton.org/news/releases/2001/cotton-genetics-research-award.cfm

The Bulletin, State University of New York College at Cortland, Issue Number 15, April 17, 2006. https://www2.cortland.edu/bulletin/issues/bulletin_05_06/April_17_06.pdf

The DuPont Challenge: Barbara Mazur. http://the-dupont-challenge.tumblr.com/post/41289108646/our-second-dupont-stem-women-all-star-is-barbara

The World Food Prize, Syngenta Scientist Dr. Mary-Dell Chilton Named 2015 National Inventors Hall of Fame Inductee, May 5, 2015. http://www.worldfoodprize.org/index.cfm/24667/35489/syngenta_scientist_dr_marydell_chilton_named_2015_national_inventors_hall_of_fame_inductee. Accessed 6 June 2015

Turning plant seeds into medicine factories, SeedQuest. http://www.seedquest.com/solutions.php?type=solution&id_article=76572, May 4, 2016

U.S. Department of Energy, Joint BioEnergy Institute, Pamela Ronald. https://www.jbei.org/people/directors/pam-ronald/

Virginia Walbot, Curriculum Vitae. http://web.stanford.edu/~walbot/cv/cv_walbot.pdf

Virginia Walbot, Professor of Biology. https://profiles.stanford.edu/virginia-walbot

(1999) A dictionary of scientists. Oxford University Press: Oxford

Authors Biography

Jill S. Tietjen, P.E. entered the University of Virginia in the Fall of 1972 (the third year that women were admitted as undergraduates—under court order) intending to be a mathematics major. But midway through her first semester, she found engineering and made all of the arrangements necessary to transfer. In 1976, she graduated with a B.S. in Applied Mathematics (minor in Electrical Engineering) (Tau Beta Pi, Virginia Alpha) and went to work in the electric utility industry.

Galvanized by the fact that no one, not even her Ph.D. engineer father, had encouraged her to pursue an engineering education and that only after her graduation did she discover that her degree was not ABET-accredited, she joined the Society of Women Engineers and for over 35 years has worked to encourage young women to pursue science, technology, engineering and mathematics (STEM) careers. In 1982, she became licensed as a professional engineer in Colorado.

Tietjen starting working jigsaw puzzles at age two and has always loved to solve problems. She derives tremendous satisfaction seeing the result of her work—the electricity product that is so reliable that most Americans just take its provision for granted. Flying at night and seeing the lights below, she knows that she had a hand in this infrastructure miracle. An expert witness, she works to plan new power plants.

Her efforts to nominate women for awards began in SWE and have progressed to her acknowledgement as one of the top nominators of women in the country. Her nominees have received the National Medal of Technology and the Kate Gleason Medal; they have been inducted into the National Women's Hall and Fame and state Halls including Colorado, Maryland and Delaware; and have received university and professional society recognition. Tietjen believes that it is imperative to nominate women for awards—for the role modeling and knowledge of women's accomplishments that it provides for the youth of our country.

Tietjen received her MBA from the University of North Carolina at Charlotte. She has been the recipient of many awards including the Distinguished Service Award from SWE (of which she has been named a Fellow), the Distinguished Alumna Award from the University of Virginia, and she has been inducted into the Colorado Women's Hall of Fame. Tietjen sits on the boards of Georgia Transmission Corporation and Merrick & Company—of which she is Vice Chair. Her publications include the bestselling and award-winning book *Her Story: A Timeline of the Women Who Changed America* for which she received the Daughters of the American Revolution History Award Medal.

She is delighted to be collaborating with her sister, Laura Privalle.

Laura Privalle received her B. S. in biochemistry from Virginia Tech. Since Jill went to UVA and our parents insisted their four kids stay in-state, Virginia Tech was the obvious choice. During freshman orientation, her father (liberated by Jill's choices) suggested that she not settle for a Master's degree but consider getting a Ph.D. Of course, he would have preferred that she select engineering as her field of study and even on the day she received her diploma, he asked when she was going to switch to chemical engineering! Laura received a M.S. in botany from Virginia Tech where her thesis work was on cellulases from *Achlya bisexualis*, a Phycomycete. She then entered the University of Wisconsin where she received her Ph.D. in biochemistry in 1983. Her major professor was Dr. Robert H. Burris and her dissertation was on nitrogen fixation in *Anabaena* 7120, a cyanobacterium. After a brief post-doctoral fellow position at Duke University working on spinach nitrite reductase, she joined Ciba-Geigy in 1984 as a post-doctoral fellow working on the control of ethylene biosynthesis and became permanent in 1986. Projects at Ciba-Geigy included herbicide detoxification, nitrogen utilization and insect control. In 1992, she joined Regulatory and Government Affairs, tasked to build the Regulatory Science group and was deeply involved in the safety assessment of Bt 176 maize, the first Bt maize product to receive regulatory approval. In 2003, Laura joined BASF to build their regulatory science group where she stayed until 2013 when she joined Bayer as the Global Head, Regulatory Field Studies.

Making the transition from academic research to more applied work and then finally to Regulatory Science allowed Laura to indulge in her passion for the acceptance of biotechnology as a critical component of sustainable agriculture. All her siblings (as well as her kids and nephews) attended Nature Camp for two weeks every summer from the time they had finished 5th grade until after 10th or 11th grades. This was a camp sponsored by the Virginia Garden Clubs with conservation as its theme. There the conservation pledge was recited daily ("I give my pledge as an American to save and faithfully defend from waste the natural resources of my country …"). Camp consisted of classes outside on geology, limnology, botany, herpetology (the favorite of boys), ornithology, astronomy, entomology, and more. This was the highlight of every year and fueled her desire to become a plant scientist and impacted many of her ensuing career decisions. It also helped that her mother was heavy into recycling long before it became common practice (1960s–1970s). Understanding plant biochemistry and being able to relate it and its value to her family certainly pushed her towards a career in biotechnology. Of course, timing is everything. She became employable just at the time when agricultural biotechnology was taking off.

During her 25 years in the Agricultural Biotechnology Regulatory Science area, Laura has participated on many intra-industry organizations as well as global bodies tasked with ensuring the safety of these products. This has led to interactions with regulators from around the world at workshops designed to gather the best scientific minds to consider the most appropriate way to ensure the safety of our food supply. She has gotten to see first-hand how countries interact to make decisions on the global food supply by sitting in on a Codex Alimentarius Commission Ad

Hoc Committee on Allergenicity meeting, She has participated in Scientific Advisory Panel Meetings of the US EPA and has attended workshops at the OECD (Organization for Economic Cooperation and Development), EFSA (European Food Safety Authority), etc. She has participated in workshops on biotechnology regulations in Japan, The Philippines, Korea, China, India, Argentina and Brazil. She has about 50 publications and 4 patents.

Agrobacterium. A Memoir (In Part Reprinted from Plant Physiology Vol. 125, 2001)

Mary-Dell Chilton

Prologue

It was very safe and comfortable being the wife of an Assistant Professor of Chemistry at the University of Washington in Seattle. Being a postdoc in the Department of Biochemistry at the University of Washington was also comfortable. My place in the academic scene was clear. The discomfort came when the postdoc ended, my second baby arrived, and there I was, a straight A student with state of the art training in DNA manipulation, wishing that I had interesting work but not wanting to miss this fleeting time when my children were small. I discussed the situation with my husband and sometime collaborator Scott Chilton, a solid gold husband and father if ever there was one, and he kindly and repeatedly gave the same counsel: "Do what you want. You can decide." My qualifications looked promising. There were no longer nepotism rules. But the problem was that I lacked geographical mobility. We loved Seattle, and Scott was advancing in his academic career. We had no desire to move. If my scope had been the entire country, at that time surely I could have found a suitable academic appointment that would allow me to work on DNA. In the 1970s, this was such an arcane interest that there was no hope of pursuing it outside of academia. Besides, I wanted to teach. I wanted to do research. And I wanted to spend time with my children. What should I do? I am human. I dealt with the issue by dithering.

M.-D. Chilton (✉)
Syngenta Biotechnology Inc., P.O. Box 12257, Research Triangle Park, NC 27709, USA
e-mail: marydell.chilton@syngenta.com

© Springer International Publishing AG 2017
L.S. Privalle (ed.), *Women in Sustainable Agriculture and Food Biotechnology*,
Women in Engineering and Science, DOI 10.1007/978-3-319-52201-2_2

A Promising Call

It was a chilly wet December day in Seattle. The year was 1970, which I mark by the birth of my second baby, Mark, who would grow to become the best friend of his two year old brother, Andrew. The phone was ringing insistently as I, with infant Mark in one arm, sprinted to catch what turned out to be the most important call of my career. It was Prof. Helen R. Whiteley, Department of Microbiology and Immunology, University of Washington, Seattle. She had been given a teaching assignment that she would like to pass along to me. It would be part time. Would I be interested? (Interested? I would be ecstatic!) The course was a laboratory on methods in DNA manipulation. This job would be a temporary instructor's position, with no hope of extension. Professionally it was not what I wanted, but at least it would get me out of the house. This job would not be a millstone, I told myself. It would be a stepping stone. I would have teaching experience! I said yes immediately. It was not necessary to discuss it with Scott, who would predictably say, "Do whatever you want!"

How I Met Agrobacterium

I taught the entering class of graduate students much of the DNA methodology I had learned over the previous three years as a postdoctoral fellow. As a final exercise, I had each student present, for the class to evaluate, a paper from the recent literature that employed one or more of the techniques that they had learned in my course. Tom Currier presented a paper from the State University of Leiden in the Netherlands, authored by Robb Schilperoort and collaborators. The paper was intended to test an astounding model for how *Agrobacterium* causes crown gall tumors on plants. The bacteria supposedly transferred DNA to the plant cells, and the transferred genes supposedly triggered growth of the plant cells into a gall. The basis for this model will be discussed below. It piqued my interest because it contradicted all the genetics I had learned and my own research results: Bacteria would only incorporate donor DNA if it matched their own DNA. Mismatch of a few base pairs could be tolerated, but overall the DNA must be homologous in order for recombination to occur. *Agrobacterium* DNA in plant cells sounded like science fiction!

The Schilperoort Paper

Tom Currier described for my class the way Schilperoort et al. measured how much "hot" (labeled) *Agrobacterium* DNA "hybridized" to filter-bound crown gall tumor DNA. As positive control, they measured how much hot *Agrobacterium* DNA hybridized to filter-bound *Agrobacterium* DNA. Surprisingly, their tumor DNA filters "hybridized" with far more of the hot *Agrobacterium* DNA than their

Agrobacterium DNA filters. This result was kinetically impossible. Tom Currier had chosen an excellent paper for my class to discuss and analyze. The students readily noted the need for further controls, especially because the authors were trying to test such an astounding model of bacterial gene transfer to plant cells. My fingers were itching to work on this DNA project!

A Project Is Launched

Tom Currier told me that his advisor, Prof. Gene Nester of the Department of Microbiology and Immunology, was interested in starting a research project on *Agrobacterium*. It was not difficult to convince Nester that he could benefit from my DNA technology to test definitively the gene transfer hypothesis. We recruited another colleague, Prof. Milt Gordon of the Biochemistry Department, a plant virologist with TMV and tobacco experience, to join our project team. I wrote up the appropriate DNA experiments as a research proposal, and with modest research grant support we founded the Seattle Crown Gall Group. I had my first job. My initial job title was Assistant Biologist, the same as that of the fellow who cleaned mouse cages in the animal room, as I recall.

Amongst our earliest experiments, we repeated the DNA filter hybridization studies in the Schilperoort paper and performed additional controls: kinetics of the reaction and melting curves of the hybridization products. We found that our data did not support the idea of DNA transfer from *Agrobacterium* to the plant cells. The telling control was the use of labeled heterologous DNA. We found that these tumor DNA filters would "hybridize" with any type of labeled DNA we added. Impurities (likely polysaccharides) in the tumor DNA made the DNA-filters sticky. Better DNA purification methods would be essential.

Does Wrong Evidence Prove the Model Wrong?

At this point, it was certainly tempting to throw out the baby with the bath water. If the evidence was wrong, then the astounding conclusion must also be wrong. And, in hindsight, that would actually be the literal truth. The DNA we tested was NOT what was transferred to the plant cells. But if that had led us all to give up and go work on something different, we would have lost the baby, which, alas, had the misfortune of being several years premature. The fact that we continued to study *Agrobacterium* reflects the mix of personalities involved. In our Seattle group, we had some believers and some doubters, and they seemed to switch roles from week to week. If you were to interview the surviving members of our original group today, you would get different stories from different people about who thought what. I think they would all be true, but for different times. Each of us likely has a very human tendency to remember best the versions of our own stories that depict ourselves as prescient. For the record, I was (intellectually) dragged kicking and screaming to the conclusion that *Agrobacterium* could put DNA into plant cells. Once we had our own direct evidence, I saw the light. But it was a conclusion from careful DNA analysis, not a "belief" based on the indirect clues from earlier work.

In the year 2000, now 16 years ago, I wrote what I called a memoir about the beginnings of the *Agrobacterium* story. With your indulgence, dear reader, I prefer to reprint that part of the story rather than retelling it. I will offer my more current thoughts in the form of an Epilogue. It will not be a review of recent literature, which others are more qualified to produce. Perhaps you could liken it more to some grandmotherly thoughts from an aging, hopefully wise, experimenter who refuses to lay down her pipets because she is still having fun. Stay tuned!

Memoir

This little memoir is not a review; the reader is directed to current authoritative *Agrobacterium* reviews with genetic (Zhu et al. 2000) or cell biology emphasis (Zupan et al. 2000). Likewise, this is not an update on recent advances in plant genetic engineering, which are the subject of a recent book (Hammond et al. 1999). Rather, I invite you to join me on a foray through the story of *Agrobacterium* transformation of plant cells. Our journey will take us back in time about 30 years, and we will note early contributions from laboratories around the globe, including Belgium, the Netherlands, France, Australia, and several in the United States. The scientists in our story represented many disciplines, from traditional ones such as plant pathology, microbiology, and chemistry to younger fields such as molecular biology, plant tissue culture, and plant metabolic chemistry. Many in the course of investigating *Agrobacterium* found intellectual haven in the newly emerging field of plant molecular biology. Beginning at a time when bulk DNA was analyzed as a macro-molecule, our story spans the birthing and growth of recombinant DNA technology.

Lest the experiments we revisit seem simple when viewed from the 21st century, our first stop will be a museum of molecular biology research in the time about which I will write, circa 1970. The catalog of restriction endonucleases was unrecognizably thin. What few enzymes were available often were tainted. Kits were unknown. Procedures often did not work. We sized DNA and determined its percentage of G and C in the model E ultracentrifuge. We measured small volumes with 5-, 20-, 50-, or 100-μL glass capillaries. We cultured our plant calli in jelly jars and fleakers. Instead of laminar flow hoods we worked in still air hoods. A few years later when the plasmid came into our lives, we taught ourselves how to do gel electrophoresis, and we designed and built our own gel rigs. (The one with the agarose wicks was known, of course, as the wicked gel.) We made combs from square aluminum rod, using double stick tape to mount teeth that were pieces of glass cut with great difficulty from microscope slides. Each of us hoarded his or her own collection of glass teeth, and it was not uncommon to hear an anguished voice cry "Who took my teeth?" Research in this period presented unique challenges.

The first cloning of DNA was out of sight, just over the horizon and of course PCR was not yet conceived.

With this setting in mind, then, let us turn our attention to the crown gall problem and consider what was known at the beginning of the 1970s.

An Idea Born Before Its Time

Dr. Armin Braun of the Rockefeller University (New York), whom many regard as the godfather of the crown gall story, first demonstrated that tumor cells are transformed, i.e. they can be freed from *Agrobacteria* and grown in vitro without the supplemental auxin and cytokinin required by normal plant cells in vitro (Braun 1958). Braun kept tumor lines growing on hormone-free medium quite literally for decades. He reasoned that *Agrobacterium* must give these cells something, and he proposed that this gift must replicate because it is never lost by dilution. He proposed for it the term TIP (tumor inducing principle).

Georges Morel of the Institut National Recherche Agronomique on the grounds of the Palais de Versailles in France discovered copious amounts of new metabolites—octopine and nopaline—in cultured crown gall tumor cells that were free from bacteria (Petit et al. 1970). Morel's group showed that the *Agrobacterium* strain, not the plant, determines the opine made by the tumor. Furthermore, each *Agrobacterium* strain can grow on its own particular opine but not on a different one. He thought Braun's TIP must be or include a gene responsible for opine synthesis in the plant. He proposed that a single enzyme catalyzed opine synthesis in the plant and opine breakdown in *Agrobacterium*, in order to account for the strain specificity of opine catabolism. We now know that part of Morel's model was not correct (the bacteria use a different enzyme for catabolism), but he was certainly on the right track about opine synthesis in tumors. However, the scientific community in 1970 was far from ready to accept the notion of a bacterial gene getting into a plant cell and functioning there. More direct evidence would be needed to support such a radical idea.

Bacterial DNA in Crown Gall Tumors?

Rob Schilperoort at the State University of Leiden, the Netherlands, as part of his Ph.D. research, prepared DNA filters with crown gall DNA and found that they bound radiolabeled *Agrobacterium* DNA amazingly well. The thesis and other publications of Schilperoort (see citations in Chilton et al. 1974) were an important factor in the founding of our Seattle Crown Gall Group. Microbiologist Gene Nester, plant viral RNA biochemist Milt Gordon, and I, an organic-chemist-turned-DNA-hybridizer, all were intrigued by the idea of gene transfer to plants. We realized that we three might collaboratively do a much more definitive type of experiment to identify bacterial DNA in tumors–if it was really there! In 1971 we began our collaboration. Tom Currier, Nester's graduate student, set about giving cancer to tobacco plants using *Agrobacterium tumefaciens* strains from the American Type Culture Collection (Manassas, VA). He inoculated the bacteria into wound sites in the stems of young plants and observed the development of crown gall tumors that were to make biological history.

The first contribution of our Seattle Crown Gall Group to the problem was a negative one that showed how large a challenge lay ahead. We found that the DNA-filter results reported by Schilperoort were caused by impurities (polysaccharides) in the DNA extracted from tumor cells, and that this technique did not have the sensitivity to detect 1% bacterial DNA in model mixtures (Chilton et al. 1974). (One bacterial genome per plant cell would constitute approximately 0.1%.) We next employed DNA renaturation kinetic analysis, which tested whether a high concentration of tumor DNA ("driver DNA") could make labeled *Agrobacterium* DNA ("labeled probe") renature faster. We showed that this method was sensitive enough to detect one copy of the bacterial genome per three tumor cells, but tumor DNA did not drive our labeled probe (Chilton et al. 1974). It was a clear negative result. We recognized that this method could only detect DNA corresponding to a significant fraction of our labeled probe. The bacterial genome contains perhaps a few thousand genes, so the acceleration of renaturation by even 10 specific bacterial genes in the tumor cells (a fraction of 1% of total bacterial DNA probe) would be below the limit of detection.

Tumor-Inducing Genes Are on an Extra-Chromosomal Element

Indirect genetic evidence that *Agrobacterium* might carry a virus or plasmid with tumor-inducing genes emerged from two kinds of experiments published in 1971. Hamilton and Fall at the University of Pennsylvania (Philadelphia) discovered that strain C58, when grown at 37 °C (28 °C is optimal), lost virulence irreversibly. They proposed that tumor induction must be a plasmid- or virus-born trait because of its susceptibility to "curing" (Hamilton and Fall 1971). At the same time, plant pathologist Allen Kerr at the Waite Institute in Adelaide, South Australia, was attempting to develop a biocontrol microbe to protect plants against crown gall disease. He co-inoculated avirulent and virulent *Agrobacteria* into the same sunflower plant. When he re-isolated the "avirulent" strain from the gall, it had become virulent! This transfer of virulence suggested to Kerr the existence of an extra-chromosomal element as vector for tumor induction (Kerr 1971).

Back in Seattle, Gene Nester read these reports and became convinced that there must be a plasmid in *Agrobacterium*. He and Alice Montoya reproduced the transfer of virulence with our own strains. Bruce Watson, a student in Milt Gordon's lab, reproduced the C58 curing experiment also. (The reproduction of published claims was clearly an important activity for our group, cast as we found ourselves in the role of iconoclasts. It was essential to know what could be believed.) Nevertheless, Bruce Watson repeatedly had no luck when he looked for plasmids in *Agrobacterium* using established methods (i.e. methods that were established for small plasmids).

Fig. 1 Photograph of Ivo
Zaenen, who discovered the
Ti plasmid of *Agrobacterium*

Key Discovery in Ghent: Ti Plasmid is Gigantic

In 1974, Ivo Zaenen (Fig. 1) at the University of Ghent (Belgium) cracked the crown gall problem wide open for everyone. Working in the laboratory of Jeff Schell and Marc van Montagu, Ivo Zaenen was the first to lay eyes on the megaplasmids of *Agrobacterium*. I asked him recently how he succeeded where others had failed. He replied that at first he did not recognize what he had found. He was using alkaline sucrose gradients to look for something else: a replicating form of an *Agrobacterium* phage called PS8 (whose DNA was once claimed to be in tumor DNA). He eventually found plasmids ranging from 96×10^6 to 156 to 10^6 M_r in 11 virulent strains and not in eight avirulent strains (Zaenen et al. 1974). His publication in the prestigious *Journal of Molecular Biology* is a landmark.

When news of this discovery came to us in Seattle, it set off a flurry of experiments and launched a vigorous competition between the Seattle and Ghent groups. We quickly isolated plasmid DNA from several *Agrobacterium* strains by Zaenen's method. Both groups found that strain C58 lost a megaplasmid when grown at 37°. Transfer of virulence was mediated by transfer of a plasmid. It quickly emerged that the genes for catabolism of octopine and nopaline were located on their respective giant plasmids, which the Ghent group christened Ti (tumor-inducing) plasmids.

Is There Ti Plasmid DNA in Tumor Cells?

At last with the Ti plasmid of our *Agrobacterium* strain in hand, we felt confident that we had the right probe to look for TIP in crown gall tumors. But when we performed renaturation kinetic analysis with the whole plasmid as probe, we got the

Fig. 2 Photograph of collaborators in the "brute force" experiment that first demonstrated the presence of T-DNA in crown gall tumor DNA. *Left to right* Don Merlo, Martin Drummond, Gene Nester, Daniela Sciaky, Mary-Dell Chilton (author of this article), Alice Montoya (deceased 1989), and Milt Gordon

by now too familiar result: It was not there. Our experiment ruled out the presence of the entire plasmid, but just as before, we recognized that a few genes could be there without our noticing any kinetic change. In order to settle the issue, we decided to cut the Ti plasmid into specific fragments and test each piece by renaturation kinetic analysis.

It was a brute force experiment involving everyone in the lab (Fig. 2). In order to label our probe to maximum specific activity, Martin Drummond seized the fresh ^{32}P-dCTP the moment we received it from New England Nuclear and labeled our plasmid DNA by nick translation. Daniela Sciaky digested the labeled DNA with *Sma*I (purified by Alice Montoya—the enzyme was not for sale). Daniela and I ran the preparative gel, made an autoradiogram, and decided whether the fragments looked good enough. (Too much nicking in the nick translation reaction could lead to breakage of the largest *Sma*I fragments, which then ran too fast during electrophoresis and contaminated the smaller fragments.) If the autoradiogram looked good, we all canceled plans for the weekend: the experiment had to be completed within 48 h, before radiation damage to the DNA began to affect the kinetics. I excised the 15 resolvable plasmid bands, which were passed to Don Merlo for electroelution of DNA from the gel slices. (He used a device involving many small dialysis bags that he designed for the purpose. He called it "The Cow" for reasons that I will leave to the reader's imagination.) We set up 75 renaturation kinetic assays (5 unlabeled "driver" DNAs × 15 labeled probes) and worked around the clock to sample the reactions and assay the percentage of renatured probe DNA in

525 (7 × 75) samples. Milt held the stopwatch and called out time points. We all did whatever had to be done next. I have never experienced such completely committed teamwork in my entire career, before or since. Although it is now nearly 25 years ago, I can clearly remember the moment of truth. While calculating and plotting the results amid a sprawl of printer tape from the scintillation counter, I suddenly saw that the T-DNA (as it would soon be known) was there in the tumor cells. Labeled probes of band 3AB, and later on, triplet band 10ABC, renatured faster in the presence of tumor DNA and no other part of the plasmid did.

A reviewer of the manuscript describing our finding required that we separate the doublet 3AB and determine which fragment was in the tumor. Although initial cloning experiments were just beginning in our group, we had no idea how to clone these blunt-ended *Sma*I fragments, and we found no enzyme that would cut one member of the doublet and spare the other. In desperation I finally managed to separate fragment 3A from 3B by a heroic serial electrophoresis of 4 days duration. We found that 3B was the fragment in the plant cells, and the paper was accepted (Chilton et al. 1977). Resolution of the band 10 triplet showed us that 10C was the member in T-DNA, and when we subsequently determined the fragment map of our Ti plasmid, fragments 3B and 10C were contiguous, showing that T-DNA was a single segment of the Ti plasmid.

Where Is T-DNA and What Defines It?

By this time, genomic Southern blots had been developed and were clean enough to show T-DNA bands; renaturation kinetic analysis was a dying art that nobody mourned. The Southern blots showed recognizable intact Ti plasmid fragments and in addition "border fragments" that were different in different tumor lines, suggesting attachment of T-DNA to plant genomic DNA. By analysis of Southerns of nuclear DNA, chloroplast DNA and mitochondrial DNA, the T-DNA of several tumor lines was proven to be located in the nuclear fraction (Chilton et al. 1980; Willmitzer et al. 1980).

In 1979, I moved from the University of Washington to Washington University in St. Louis, and focused on nopaline Ti plasmids, while the founding group in Seattle continued with the octopine strain. My new group at Washington University, the Seattle group, and Patti Zambryski in the Ghent group (Fig. 3) all succeeded in cloning T-DNA fragments from tumor DNA. When we sequenced through the junctions of T-DNA and plant DNA, comparing plasmid DNA with T-DNA, we found a 25-bp imperfect direct repeat on the Ti plasmid at the edges of what is incorporated into the plant genome. These border sequences define T-DNA on the plasmid but not in the plant: they are not transferred intact to the plant cell (reviewed in Binns and Thomashow 1988).

Fig. 3 Photograph of research group in Ghent, Belgium, 1984. *Left to right* Jeff Schell, a visiting scientist from China, Marc van Montagu, Patricia Zambryski, and Ken Wang (a student)

Genetic Picture of the Ti Plasmid

vir *Genes*

Transposon mutagenesis of the Ti plasmid in Leiden, in Seattle, and in Ghent showed that all mutations affecting tumor induction mapped to a sector of approximately 42 kb, separate from T-DNA, called the virulence (*vir*) region. The *vir* genes constitute a regulon inducible by acetosyringone and other phenolics that are found in plant wound juice (Stachel et al. 1985). These compounds, directly or indirectly, affect the "antenna" protein VirA, which autophosphorylates, then phosphorylates VirG, a transcriptional activator for all of the *vir* genes.

T-DNA is excised from the Ti plasmid by endonuclease VirD2, with facilitation by VirD1 and VirC1. VirD2 nicks the bottom strand of the right border sequence after the third base and attaches to the 5' end of the nick, forming the "leading" end of the T-strand to be delivered to the plant. The details of left border scission are not clear, but VirD2 produces a similar nick there. The *vir* E2 gene encodes a single strand binding protein essential for tumor induction, that can alternatively be expressed in the plant with equal effect. The VirB operon consists of 11 open reading frames, which encode the T-DNA conduit from bacterium to plant. The structural and functional similarity of many of these to proteins involved in plasmid transfer to other bacteria has led to the view that T-DNA transfer has evolved from plasmid conjugation (reviewed in Zhu et al. 2000 and Zupan et al. 2000).

T-DNA Genes

Transposon hits in T-DNA were found to eliminate opine production or to alter tumor morphology or to have no recognizable effect at all. The morphology mutations were eventually shown to eliminate cytokinin autonomy ("rooty" tumors) or auxin autonomy ("shooty" tumors). T-DNA genes were shown to encode a two-step pathway to the plant auxin indoleacetic acid and an enzyme producing the cytokinin isopentenyladenosine 5'monophosphate (reviewed in Binns and Thomashow 1988). Most importantly, no mutation in T-DNA blocked T-DNA transfer. All of the genes affecting the process of T-DNA export to the plant cell mapped in the *vir* region. This fact would greatly simplify the disarming of T-DNA and construction of *vir* region-containing helper plasmids lacking any T-DNA.

From Pathogen to Gene Vector

In order to use the Ti plasmid as a vector, we needed a method of putting genes into T-DNA (and knocking some out, as well). In Ghent and in St. Louis, methods were developed for inserting DNA into any specific part of the Ti plasmid. The DNA to be inserted was cloned between pieces of T-DNA on a plasmid, introduced into the bacterium by conjugation or by transformation, and subjected to "forced recom-bination" (Matzke and Chilton 1981; Van Haute et al. 1983). A simpler approach to engineering T-DNA was to make a small separate T-DNA plasmid that could be manipulated directly. Although *Agrobacterium*, in nature, keeps *vir* genes and T-DNA on the same replicon, there is no requirement for this arrangement. If you place T-DNA on a separate replicon in *Agrobacterium* (a binary vector, as it is now called), the process of T-DNA transfer to the plant cell still occurs with good efficiency (De Framond et al. 1983; Hoekema et al. 1983). Thus, the T-DNA of a binary vector could be engineered directly in *Escherichia coli* and then transformed into *Agrobacterium*.

Another problem for the genetic engineer was plant regeneration. All efforts to regenerate a plant from transformed cells were initially rewarded with only rare deletion mutants that had lost practically all of their T-DNA, a strong indication that at least part of T-DNA was inimical to plant regeneration. We discovered the critical part almost serendipitously. Tony Matzke and Ken Barton, post-docs in my group, introduced a yeast gene into T-DNA in a position that we thought might hit an oncogene (Matzke and Chilton 1981). It turned out indeed to inactivate the cytokinin production gene. In collaboration with Andrew Binns at the University of Pennsylvania, we discovered that this single insertion event produced an engineered T-DNA that was completely disarmed. It produced transformants that synthesized nopaline but that could not grow autonomously without hormones. Binns identified the transformed plant cells by screening for nopaline production. In contrast to crown gall tumor cells, the tobacco cells transformed by multiple copies of this

T-DNA were able to regenerate into normal plants that passed the T-DNA copies to progeny plants as Mendelian traits (Barton et al. 1983). By 1982 we had the first evidence that foreign DNA engineered between T-DNA borders and transformed into the plant nuclear DNA could be stably maintained in the plant genome and passed intact to progeny.

Starting in about 1980, a formidable new group was assembled by Ernie Jaworski at Monsanto (our neighbor in St. Louis) to harness the T-DNA transfer technology for crop improvement. At this time Michael Bevan in my laboratory found himself in a race with Patti Zambryski's team in an effort to sequence the nopaline synthase (*nos*) gene and map its promoter and terminator by S1 nuclease protection (Bevan et al. 1983; Depicker et al. 1982). Then a second race ensued amongst Bevan, Zambryski and her Ghent collaborators, and the Monsanto group to isolate the *nos* gene promoter and splice it to a kanamycin resistance coding region in order to create a selectable marker that might work in plant cells. If this scheme worked, then one would no longer have to screen for nopaline production to find transformed plant cells: one could select the cells with T-DNA inserts on kanamycin agar.

The symbolic coming of age of genetic engineering occurred at the Miami Winter Symposium, January 18, 1983. During one session, Jeff Schell, Rob Horsch from Monsanto, and I all gave talks about *Agrobacterium* and its adaptation as a gene vector for plants. All three of us reported success with chimeric kanamycin resistance genes as a selectable marker for plant cells (Bevan et al. 1983; Fraley et al. 1983; Herrera-Estrella et al. 1983). I described initial success in transforming tobacco cells with binary vectors (which we called MiniTi at that time). In addition, I described our tobacco plants engineered with a disarmed Ti plasmid, and Southern blots proving that they passed their T-DNA insert to progeny intact. It was clear from the progress in all three groups that crop improvement by genetic engineering would become a reality.

Reflections from 2000

Finding that T-DNA can integrate into a plant genome without benefit of homology was a real intellectual shock to me. The bacterial transformation studies I had made as a student and again as postdoc taught me the absolute need for good homology in those systems. Now illegitimate recombination seems the rule not only for T-DNA but also for foreign DNA integration in animal cells and indeed naked DNA delivery to plant protoplasts or bombardment of plant cells with DNA-coated microprojectiles. Incorporation of foreign DNA is clearly a process that cells carry out efficiently, perhaps in the course of repairing genomic damage. It is not a trade secret of *Agrobacterium*, although we may yet discover some secret details of the process. *Agrobacterium* acted as an inspiration to others who have developed various means of DNA delivery. It gave us our first selectable marker (tumor induction). It gave us the promoters and terminators for the next generation of selectable markers (octopine/nopaline synthase promoters and *nos* terminator, bounded as they were by T-DNA borders and neighboring genes in this highly compact T-DNA). The wide host range of *Agrobacterium* (even wider now that monocot transformation is facile) inspired the idea that DNA incorporation may be a universal phenomenon. But perhaps the most important legacy from *Agrobacterium* has been its inspiration of confidence that foreign gene integration, even though DNA is sometimes delivered artificially, is a perfectly natural process. My most fervent wish is that well-meaning environmental proponents will come to recognize this and embrace the technology based on it.

Epilogue

Will This Really Work in Crops?

The memoir reprinted above concluded in 1983, at the beginning rather than the end of a chapter of history. We could see that the method of introducing new genes into plants was going to work. It was crude, but it could be improved. What was less clear, to me at least, was how to improve real crop plants. What pests—viruses, bacteria, fungi, viroids—are a problem for which crops and what genes would be useful to combat them? What environmental conditions affect yield in which crops? How could we find genes to make the crop tolerant to drought, flood, cold, heat, famine, plague or pestilence? Could we teach *Agrobacterium,* whose host range was known to include dicots and gymnosperms, how to deliver genes to monocots such as corn, wheat, rice and other cereals, the most important food crops? I never even considered the idea that the public might fear these improved plants! All we had at that point was a sound and promising start.

A Visitation

One additional aspect was not clear to me: what role would I have in harnessing this new technology. I realized that if I were to carry on, I would need to develop a strong collaboration with scientists experienced in agriculture. I remained a consultant for the genetic engineering group at Monsanto, and their support of my university research program continued, but they seemed to show little interest in my longer term career. This matter seemed to be of greater interest to three businessmen from Swiss multinational CIBA-Geigy, who visited me at Washington University in mid-1982. They disclosed, after a lengthy discussion, that they had been tasked with identifying a leader for their projected new agricultural biotechnology group to be established in North Carolina. They wanted to know whether I would be interested in being a candidate. Suppressing my excitement was difficult. I asked for some time to think it over. Scott, as was his custom, advised: do whatever you want! Both sets of aging parents would be nearby. Scott was from Virginia and I grew up in North Carolina.

CIBA-Geigy, which today after two mergers, spinoffs and name changes is called Syngenta, offered me the position in early 1983, and we moved to Raleigh in the summer of that year. It was a real adventure for each of us. Scott had the challenge of finding a new academic position, and this time it was he who lacked geographical mobility. He turned out to be quite successful at this, and before long the Botany Department at North Carolina State University was both happy and fortunate to hire him as a visiting professor. It turned out to be a lengthy visit.

Beginnings

Meanwhile, I had many new responsibilities but was given plenty of help. I had both a US boss and a Swiss boss, and our operation was managed by the Biotech Executive Committee, consisting of the three of us plus two additional Swiss executives. We had to invent the Agricultural Biotechnology Research Unit (ABRU) from the ground up. Working in leased facilities for the first year and a half while building our own new laboratory in Research Triangle Park, N. C., we developed a project portfolio and began recruiting scientists with relevant skills. CIBA-Geigy was in part an agricultural chemical business, and many of the chemists were not pleased at the prospect of biotechnology solving problems that they viewed as their purview. The KL (top Swiss Executive Committee) subscribed to the longer term view that if biotech could capture in part the value of the agricultural chemicals business, it was prudent for CIBA-Geigy to enter this new biotechnology business. Our Biotech Executive Committee had to decide how to do that. Would we enter the seeds business more extensively, or would we license biotech to others, or both?

The Overall Lesson

Choosing project objectives was a very different process from what I had used in the university setting. We had an expert from Basel headquarters who helped us to figure out, for each project idea, what financial value would be added to the genetically modified seed by our new gene(s), assuming that the science worked well. What would the improved seed save the grower in input cost? What would be the value of any increased yield (or prevented loss of yield)? Would the yield have higher intrinsic value because of nutritional improvement, improved processing attributes, etc.? Could it be segregated from bulk crop yield and sold at higher price, or perhaps at the same price but ballooning market share? Next, my scientists and I estimated how long the project would take and what was the probability that it would work as hoped. Additional experts were consulted to consider other business aspects—were we in the right business with the appropriate customer base, in the right geographical region and country? Were there regulations in place for testing and eventually selling our genetically modified seeds (and for the grower to sell his crop, of course)? And then there was the vital question of whether our product could be blocked from the market by patents of competitors. The overall lesson of this exercise was that the seed business, and even more so the genetically modified seed business, was going to be a challenge.

The Big Picture

Since the dawn of the technology in January 1983 (date of the Miami Winter Symposium), 33 years have now passed. *Agrobacterium* has kept its promise. The bacteria deliver DNA to monocots, soybeans, and (I suppose) any plant it wants, or indeed we want, not to mention yeast, fungi and probably robots and iPads. Inspired by this clever microbe no doubt, additional methods of DNA delivery have been perfected. While these methods scatter donor DNA to random locations on host chromosomes, newer methods have been developed for directing the donor DNA to a desired location (gene targeting), and those same methods can be used to edit the plant (or other) genome quite precisely. Technology for manipulating DNA available today is mind-boggling!

In 1983 the main limitation was technology. The technology that my postdocs Tony Matzke and Ken Barton and I developed in collaboration with Andrew Binns, primitive though it appears today, was prized at that time because it provided the proof of concept. At that time, the company that could transform the highest value crop with greatest efficiency promised to be the biggest winner. Currently that aspect is not such a problem and the big challenge is to identify the best genes to address the plant's problems. We are clever and I believe we can do that. But today we face a challenge that goes beyond the science: the societal part of the picture, which had hardly crossed our minds in 1983: The product must be attractive to the consumer or at least to the end-user (who must convince the consumer of its value), if genetically modified seed is to sell.

You will read about many of the advances as you enjoy the stories in the chapters assembled here. This book will contain a collection of personal histories as much as the story of scientific advances in agriculture. I hope that it may project for the reader more than what we and others have done, but also how and why we came to do it, a part of the story that is usually interesting but rarely told.

The Message

If the prophets of global warming are correct (as I fear they are), climate change in addition to population growth will bring an urgent need for accelerating the rate of plant breeding advances. With changing climate, our food and fiber crops will face new challenges. New weather patterns will bring new insect pests that damage plants, and pests that act as carriers of new diseases. The task of the plant breeder will be even more challenging than it is today. Breeders will need the best technology available, and genetic modification of plants promises to be an important tool. It is a safe procedure that, like the traditional plant breeding of the past, we have learned from nature.

Therefore I need to leave you with one closing message, and you will not be surprised to note that it bears a strong resemblance to my plea of 16 years ago. I hope to see the technology for producing genetically modified plants accepted, even embraced, by the public in my lifetime. We must hasten if we are to succeed at this, for I am not getting any younger. If you agree with my concerns, do not keep it a secret. Spread the word when you have an opportunity. Believe me, I will do my part by continuing as long as I am able.

References

Barton KA, Binns AN, Matzke AJM, Chilton M-D (1983) Cell 32:1033–1043
Bevan MW, Barnes WM, Chilton M-D (1983a) Nucleic Acids Res 11:369–385
Bevan MW, Flavell RB, Chilton M-D (1983b) Nature 304:184–187
Binns AN, Thomashow MF (1988) Ann Rev Microbiol 42:575–606
Braun AC (1958) Proc Natl Acad Sci USA 44:344–359
Chilton M-D, Currier TC, Farrand SK, Bendich AJ, Gordon MP, Nester EW (1974) Proc Natl Acad Sci USA 71:3673–3676
Chilton M-D, Drummond MH, Merlo DJ, Sciaky D, Montoya AL, Gordon MP, Nester EW (1977) Cell 11:263–271
Chilton M-D, Saiki RK, Yadav N, Gordon MP, Quetier F (1980) Proc Natl Acad Sci USA 77:4060–4064
De Framond AJ, Barton KA, Chilton M-D (1983) Biotechnology 1:262–269
Depicker A, Stachel S, Dhaese P, Zambryski P, Goodman HM (1982) J Mol Appl Genet 1:561–573
Fraley RT, Rogers SG, Horsch RB, Sanders PR, Flick JS, Adams SP, Bittner ML, Brand LA, Fink CL, Fry JS, Galluppi GR, Goldberg SB, Hoffmann NL, Woo SC (1983) Proc Natl Acad Sci USA 80:4803–4807
Hamilton RH, Fall MZ (1971) Experentia 27:229–230
Hammond J, McGarvey P, Yusibov V (eds) (1999) Plant Biotechnology. Springer, Berlin
Herrera-Estrella L, De Block M, Messens E, Hernalsteens J-P, Van Montagu M, Schell J (1983) EMBO J 2:987–995
Hoekema A, Hirsch PR, Hooykaas PJJ, Schilperoort RA (1983) Nature 303:179–180
Kerr A (1971) Physiol Plant Pathol 1:241–246
Matzke AJ, Chilton M-D (1981) J Mol Appl Genet 1:39–49
Petit A, Delhaye S, Tempé J, Morel G (1970) Physiol Veg 8:205–213

Stachel SE, Messens E, Van Montagu M, Zambryski P (1985) Nature 318:624–629

Van Haute E, Joos H, Maes M, Warren G, Van Montagu M, Schell J (1983) EMBO J 2:411–417

Willmitzer L, De Beuckeleer M, Lemmers M, Van Montagu M, Schell J (1980) Nature 287:359–361

Zaenen I, Van Larebeke N, Teuchy H, Van Montagu M, Schell J (1974) J Mol Biol 86:109–127

Zhu J, Oger PM, Schrammeijer B, Hooykaas PJJ, Farrand SK, Winans SC (2000) J Bacteriol 182:3885–3895

Zupan J, Muth TR, Draper O, Zambryski P (2000) Plant J 23:11–28

From Maize Transposons to the GMO Wars

Nina Fedoroff

Early Days: Discovering Plant Genetics and Developing Molecular Methods for Plants

Though I didn't know it at the time, my involvement with plant biotechnology began in 1976 when I encountered Barbara McClintock by chance while visiting the Cold Spring Harbor Laboratory to give a talk. I was heading out of the Demerec Laboratory on my way to meet with then Director Jim Watson when I ran into a tiny, elderly woman. She stopped me to apologize for missing my seminar—I quickly guessed this must be the legendary Barbara McClintock, though I couldn't understand why she should be apologizing to me, as I was just a post-doc. She invited me to continue the conversation in her laboratory. On impulse, I shrugged off the meeting with Watson and followed her down to the ground floor. My recollection is that we ate peanuts and drank Cinzano—she later assured me that only the part about the peanuts was true. We talked about the usual stuff—science, science politics, a bit of philosophy.

I was puzzled. McClintock's reputation for impenetrability didn't fit with the lucidity of her casual discourse. And I was intrigued. I went back home to Baltimore, where I was a post-doc in Don Brown's laboratory at the Embryology Department of the Carnegie Institution of Washington (now the Carnegie Institution for Science), curious to know more. Since McClintock was a member of the same institution, although by then retired, she had published in the Carnegie Yearbooks, all shelved in the department's main reception area. I got them down, copied (xeroxed, in the terminology of the time) her chapters, and began to read.

It was tough going. I was a molecular biologist. Indeed, my post-doc was devoted to DNA sequencing and I'd sequenced some of the first genes ever sequenced, the 5S

N. Fedoroff (✉)
OFW Law, Washington, DC, USA
e-mail: nvf1@psu.edu

© Springer International Publishing AG 2017   39
L.S. Privalle (ed.), *Women in Sustainable Agriculture and Food Biotechnology*,
Women in Engineering and Science, DOI 10.1007/978-3-319-52201-2_3

ribosomal RNA genes of South African clawed toad, *Xenopus laevis* (Miller et al. 1978; Fedoroff and Brown 1978). I'd done some simple bacterial and bacteriophage genetics, but I knew nothing about plants, much less plant genetics. Yet even if I'd had a classical genetics background, it wouldn't have helped me much. I was into genes and promoters. Publishing 30 years earlier, McClintock wrote about a position on maize chromosome 9 where breakage could occur reproducibly during development—one that could also move to a new place on the chromosome (McClintock 1987). She called it a "transposable genetic element."

By the late 70s, insertion sequences and transposons had already surfaced in bacteria and had at least a genetic identity, although their DNA sequences were still in the future. The more I read about maize transposable elements—slowly and painstakingly—the more interesting I found them. Yes, maize transposable elements could break chromosomes, but they could also jump into genes. And they could take over the regulation of a gene, making it dependent on the transposon for its expression. Or the insertion could regulate gene expression in exactly the opposite way, the insertion silencing the gene only in the presence of a related mobile sequence. Maize transposable elements could go silent and invisible, not moving for generations, then come back to an active form. And they could talk to each other, collaborating to determine how several different genes with insertions were expressed during the plant's development.

I was approaching the end of my post-doctoral tenure and was beginning to think about what I would devote my research to in the future, assuming I could land an academic post in a research institution. I kept thinking about working on the molecular biology of the maize transposable elements—and repeatedly dismissing the idea. Plant biology didn't get much in the way of public research funding and plant molecular biology didn't yet exist. The notion that I could become a plant biologist, invent the necessary technology for cloning and analyzing plants genes, and find financial support, while starting an academic career was daunting. And I was a single parent who took her parenting seriously, so I'd pretty much shelved the idea.

Quite unexpectedly, Don Brown, by then the Embryology Department's director, called me into his office one day to offer me the staff position newly vacated by the departure of Igor Dawid. Although it had previously been announced that internal candidates would not be considered for the position, Don explained that he and his colleagues had come to the conclusion that I was being discriminated against me just because I was an internal candidate. So if I wanted the job, it was mine—elegant job offer, no?

My immediate reaction was to dislodge my tucked-away fantasy of working on the maize elements so elegantly described in McClintock's genetic experiments. Staying at Carnegie would let me to devote all of my time and effort to research, so I could gamble on solving the problems that stood in the way of molecular approaches in plants. These were not insignificant. No plant genes had yet been cloned and people were saying that they couldn't be cloned.

But first I had to persuade Barbara, who didn't have much patience for slow learners, that I was serious enough for her to teach and to share some of her precious mutant lines. I jumped right in, arranging to grow corn the following summer at the Brookhaven National Laboratories through the generosity of maize geneticist Ben Burr. Barbara did indeed give me a bit of everything she'd worked on and I probably overplanted. When pollination season came, I settled into Brookhaven housing with my son, whom I'd found a spot for in a local camp.

That summer turned into a nightmare. There's no stopping during pollination season and you have to have a pretty good idea of what you want to do—something I didn't yet have, of course, so I probably did twice as many crosses as I needed to do. My objectives and Barbara's were often at odds—her approach was that of a geneticist, while mine was that of a molecular biologist. In particular, Barbara had capitalized on the extraordinary properties of the maize kernel and its pigment-producing aleurone layer to follow the effect of the mobile elements on multiple genes simultaneously. Many of the genes she worked with coded for enzymes and—as we eventually learned—regulatory proteins involved in the synthesis and modification of the anthocyanin pigment produced by the kernel's aleurone layer. It struck me as unlikely that these would be sufficiently abundant proteins so that we could easily purify them and raise antibodies to them, the only tools then available for getting to the mRNAs encoding them and in turn to the genes. So I focused on the few genes she'd identified with insertions that were likely to encode relatively abundant proteins, such as those involved in sugar and starch biosynthesis, major biochemical activities of corn kernels.

And then there were the 5 am to 2 am days—every day! My technician Jeff had come with me to help with pollination, but it quickly became apparent that he was violently allergic to corn pollen. He was swelling up dreadfully and I sent him home to Baltimore. So all the planning of crosses, silk-trimming, tassel-bagging and pollination fell to me. I was barely managing and Barbara and I couldn't agree on anything. The summer's nadir came when my son's camp called and told me my son had impetigo—a highly contagious skin disease. I was to come and get him immediately and not bring him back for the rest of the summer.

In the end, however many wrong crosses were made, some of the right crosses were made, as well. My young son and I harvested that first crop over several brilliantly sunny warm September days. I remember marveling over each shiny, colorful ear as it emerged from is brown and shriveled husk. My knowledge of what I was looking at was improving and I was beginning to connect Barbara's infer-ences with the pigmentation patterns I was seeing—I still remember my constant amazement that she had figured out so much from such subtle clues. Jerry Neuffer once told me that every time he thought of a clever experiment to do with maize transposons, he found that Barbara had already done it years earlier. I felt the same way for a long time—my first single-authored purely genetic paper was still a decade in the future.

In the meantime, there was biochemistry to be done in order to get a toe-hold in the molecular realm on the way to cloning maize genes with McClintock trans-posable element insertions. My group was growing and post-doctoral fellows Susan Wessler and Mavis Shure took on the challenge of cloning a cDNA copy of an mRNA that looked promising both because it was likely to encode an abundant protein and because it was produced from a gene that carried a complete McClintock-characterized transposon.

The gene we chose was the *Waxy* gene, the wildtype version of which was already known to encode a starch granule-bound UDP-glucose starch transferase responsible for the synthesis of amylose in the kernels. The fact that it was bound to starch granules and that we had both wildtype and mutant maize varieties made it relatively easy to identify an abundant 58-kD starch granule-bound protein that was present in wildtype kernels, but not in null *waxy* mutants. Though we were never able to solubilize an active protein, the genetic evidence that we'd identified the protein encoded by the *Waxy* gene was strong enough for us to raise antibodies to the protein and identify a cDNA clone using a now all-but-forgotten technique called "hybrid-selected translation."

McClintock's genetic treasury comprised basically two kinds of insertion mutations, ones in which the inserted transposon was itself mobile and others in which the insertion was mobile only in the presence of a second, mobilizing element. I named these "autonomous" and "non-autonomous" elements, respectively. These, in turn, fell into families according to which non-autonomous elements were mobilized by a given autonomous element. My first hypothesis was that the non-autonomous elements of a family were likely to be mutant elements lacking an active version of gene that encoded its "transposase," an enzyme postulated to cut and resect DNA to move the transposon to a new site. In the first transposon family that Barbara had identified, the *Activator-Dissociation* family, the *Activator* (*Ac*) element could transpose autonomously, while the *Dissociation* (*Ds*) element, first named for its ability to "break" or "dissociate" a chromosome, could only move in the presence of an *Ac* element.

I chose the Waxy locus for one of the first gene cloning efforts because McClintock had identified both an *Ac* insertion and a *Ds* insertion at the locus. Moreover, the *Ds* line appeared to have arisen from the *Ac* line, so it was a fair guess that it was a non-mobile mutant derivative of the parent *Ac*. Though it was slow going, the molecular techniques developed for other kinds of organisms worked relatively well after a bit of tweaking for Wessler and Shure to clone a cDNA copy of the mRNA encoded by the Waxy gene (Shure et al. 1983). So it should have been perfectly straight-forward to clone the gene. But it wasn't. A third post-doctoral fellow, Debbie Chaleff, had worked long and hard to clone maize DNA into the lambda vectors that were widely used at the time. She had failure after failure and finally left the lab in frustration.

And indeed, there were rumors in the plant community that plant DNA couldn't be cloned. I refused to believe them. Confident that all that was required was extreme care and attention to the success of each step in the process, I took on the project myself. I worked meticulously, devising controls for every step in the process—more controls than anyone ever used. But I, too, failed to clone maize DNA in the then widely-used lambda vectors. And I failed again. Then, for reasons I cannot reconstruct, I grew overnight cultures of all the bacterial strains in my collection and plated the packaged, recombinant, maize-DNA containing viruses on all of them. Amazingly, one strain—I even remember it was designated

K803—showed hundreds of plaques. I plated it again, and again had many plaques. The problem was solved.

Some time later, I remember receiving a call from a researcher at the Rockefeller University (if my memory serves me correctly) during this time, telling me that he had simply cried in frustration over his failure to clone maize DNA. I sent him some K803, which I had received from a laboratory at Cold Spring Harbor, and it worked for him as well. While I never stopped long enough to nail down the reasons for the earlier failure, I did carry out one experiment that strongly suggested to me that the trouble resided in the higher levels of methylation of plant DNA compared with animal, lower eukaryotic and bacterial DNAs. And that was simply to collect the virus particles grown in the K803 strain and ask whether they showed similar titers on both K803 and the other strains people were commonly using as bacterial hosts for recombinant lambda viruses—they did. I published a short description titled "Notes on cloning maize DNA" in the Maize Genetics Newsletter in 1983 in the hopes of helping others get past this cloning bottleneck (Fedoroff 1983a).

Explaining and Cloning Maize Transposons

During the period that my laboratory was struggling to develop biochemical and molecular methods for plants, Barbara was asked to write a chapter for a book that Jim Shapiro was organizing titled "Mobile Genetic Elements." During one of my not infrequent visits to her laboratory during this time, Barbara asked me to co-author the chapter with her. She was quite frustrated with her inability to communicate her science to others and blamed it on her lack of writing ability. She eventually asked me to write the chapter myself. I was not yet confident that I understood all the subtleties of the maize elements and their genetic and epigenetic behaviors, but I agreed to take on the task, assuming that she would set me straight if I got any of it wrong.

I had, by then, read and reread her papers several times and I had been through several seasons of genetic crosses. I invested a good deal of time in taking pictures of kernels from what were now my own maize stock that illustrated the pigmentation and variegation patterns McClintock used so effectively to understand the genetic behaviors of maize transposons. I wrote the draft painstakingly from a molecular perspective. Since the chapter was written before the first maize transposable elements had been cloned and analyzed in my laboratory, I relied on my knowledge of the literature describing prokaryotic transposons to interpret McClintock's genetic studies in a way that would be comprehensible to a molecular geneticist of my generation. I sent off a draft to Barbara with a request that she criticize it mercilessly. I heard nothing. When I at last summoned up the courage to call her, she declined to comment on the manuscript. When I queried her on her reasons, she said that if she commented on it, she would have to take responsibility for it—and she refused to do so.

I was stunned. I didn't think I could have made such a complete mess of it for her to refuse any further discussion—but that was, of course, a possibility. I knew that I had used very different terminology from that she'd invented so many years earlier when people didn't have words for what she was seeing in her genetic experiments. And I had not adopted her favored hypothesis that these were misplaced regulatory elements: she had come to call them "controlling elements." (That said, I can say with all the clarity of hindsight that she was not far off the mark, as the kinds of regulatory interactions among the elements that she first identified and we later characterized at the molecular level are very consistent with the regulatory interactions among non-mobile genes. And more than that, bits of transposons are often in the regulatory sequences of genes.)

So I guessed that she was reacting badly to my rather different treatment of her intellectual construct—and decided that I had to move ahead with the manuscript. It was perhaps the most painful piece of writing I have ever done, agonizing over each inference, checking it again and again, sentence by sentence. In the end, Barbara made peace with my writing about her elements and even suggested that I'd contributed to her selection as a Nobel laureate. Though I do not believe that was true, I didn't get much wrong in that early piece, judging from the molecular evidence the accumulated over the following decades. And I do believe that book chapter, which appeared in 1983, became an accessible entry point for students to understand McClintock's work in a contemporary context (Fedoroff 1983b).

Once I'd solved the problem of cloning maize DNA, the research moved very quickly. By early 1983, the year McClintock received an unshared Nobel prize, we wrote the first paper describing the cloning of the *Waxy* gene (Shure et al. 1983) and the first *Ac* and *Ds* elements (Fedoroff et al. 1983). I had cloned the *Waxy* gene from both a wildtype revertant strain of what Barbara designated the *Ac wx-m9* allele and two strains with *Ds* insertions in the *Waxy* gene. One of these was designated *wx-m9* by McClintock and was a direct derivative of the *Ac wx-m9* allele, while the second, *wx-m6*, was of independent origin and arose by transposition of a *Ds* from another position on the same chromosome. When the wild-type and *Ac wx-m9* recombinant viral DNAs were denatured, re-annealed, and examined by electron microscopy, I saw a double-stranded molecules with a long, single-stranded loop, confirming the long-standing conjecture that the *Ac* transposon was a piece of DNA, in this case about 5 kb in length, inserted precisely into the gene. And just as I'd guessed, that first *Ds* turned out to be an *Ac* transposon with an internal deletion. The second *Ds* element, cloned from the *wx-m6* allele, contained a shorter element comprising about a kilobase from each end of the *Ac* element. It, too, likely arose by an internal deletion, although it had moved into the *Waxy* gene as an already mutant, non-autonomous *Ds* transposon.

Gene Tagging with Maize Transposons

An observation that grew out of analyzing maize DNA with the first three transposons of the *Ac–Ds* family eventually gave rise to the active transposon subfield of "tagging" using maize transposons in a number of different plants. Here's how it happened (Fedoroff et al. 1983). I used the 2-kb *wx-m6 Ds* as a probe for transposon ends and a short fragment from the middle of the *Ac wx-m9* transposon as a probe for the central part of the transposon. Importantly, the selected fragment covered the deletion site in the *Ds* element cloned from the *wx-m9* allele and therefore gave a shorter fragment. When DNA from either the *Ac wx-m9* or *wx-m9* strains were probed with the *wx-m6 Ds* probe homologous to the ends of the *Ac* transposon, I picked up many homologous fragments of many different sizes. Yet when I probed the respective maize DNAs with a probe corresponding to the middle of the element, I picked up fewer bands, only one of which was the exact size of the fragment in the cloned *Ac* element. There was also a fragment in that size range in the *wx-m9* DNA and it co-migrated with the central fragment cut from the *Ds* transposon cloned from the *wx-m9* DNA. That meant very simply that the genetic observation that there was a single *Ac* element in the genome carrying the *Ac wx-m9* allele and it was the one inserted in the *Waxy* gene!

Thus the single genetically active *Ac* element appeared to have a unique structure. The very obvious implication was that it might be possible to identify a gene in which there was an Ac insertion mutation without knowing anything about the gene product. In view of how much we know about gene structure, homology and function today, it is difficult to imagine how hard it was back then to clone a gene knowing nothing about it other than its genetic behavior. Since that was decidedly the rate-limiting step in gene isolation at the time, we decided to determine whether we could use what we knew about *Ac* structure to clone a gene whose gene product was far less abundant than that of the *Waxy* locus. We picked the *Bronze (Bz)* gene, one of the genes in the anthocyanin pigment biosynthetic pathway and one that had figured prominently in McClintock's early analyses of maize transposon behavior.

The *Bz* gene encodes a UDPglucose-flavonol glucosyltransferase whose activity stabilizes the deep purple anthocyanin pigments produced in the kernel aleurone layer. It is not an abundant enzyme and efforts to purify it directly had not gone well. But McClintock's *Ac bz-m2* allele carried an intact, autonomous *Ac* transposon and she had identified a derivative, *bz-m2(DI)*, that carried a non-autonomous *Ds* element. It had been derived from the parent *Ac bz-m2* allele and was likely to have a mutated *Ac*, much like the one I'd characterized at the *Wx* locus. We screened recombinant viruses containing fragments of DNA cut with enzymes that we knew didn't cut within the *Ac* sequence with a short probe corresponding to the center of the *Ac* element. We isolated 25 such clones; of these, 6 gave fragments of the right size for an intact *Ac* element and 4 of these 6 had the same flanking sequences, which turned out to be the *Bz* locus, as judged the comparative sizes of homologous fragments in strains with and without *Ac* insertions (Fedoroff et al. 1984).

While a number of investigators went on to use precisely these observations to clone genes from maize strains with insertions of both *Ac* and other transposons, we thought that this "transposon tagging" methodology would be even more widely useful if we could show that maize transposons could move in plants other than maize. The first step was to show that *Ac* could move in another plant and the obvious choice, for ease of transformability, was tobacco. I had raised the possibility of a collaboration during a phone call with Jeff Schell, than already Director of the Max Planck Institute in Cologne, whose laboratory was carrying out many of the pioneering studies on the plasmids of *Agrobacterium tumefaciens*. The following summer I visited the Max Planck Institute and, I began a collaboration with Barbara Baker in Schell's group.

To function in another plant, the *Ac–Ds* system necessitated correct expression of the element-encoded proteins as well as complementation by other proteins required for transposition. The first effort was quite straightforward and involved transforming the original *Ac* and *Ds* clones from the *Wx* gene into tobacco cells using *A. tumefaciens* Ti-plasmid vectors. The transformed clones were probed with both transposon probes and a probe to detect the short fragment of *Wx* DNA surrounding the insertion site. The results were positive: the empty donor site fragment was detectable in the lines transformed with the *Ac* element, but not those transformed with the *Ds* element (Baker et al. 1986). In a subsequent publication, the detection system grew more sophisticated (Baker et al. 1987), and in the ensuing years, marked transposons based on the *Ac–Ds* family and later the *Suppressor-mutator* (*Spm*) were widely used. *Ac–Ds* family transposons were even shown to transpose in organisms as varied as *Arabidopsis* and carrots (Van Sluys et al. 1987), to rice (Qu et al. 2009), tomato (Levy et al. 2000), yeast (Weil and Kunze 2000), zebrafish, and even human cells (Parinov et al. 2006).

Fortunately, I had filed a patent application on using transposable element to clone plant genes, and the patent was granted in due time. Over the years that it was in force, the patent brought in substantial licensing fees from companies, some of which have been committed to funding (upon my demise) a post-doctoral fellowship honoring McClintock within the Carnegie Institution for Science, home institution to both Barbara McClintock and me for a good fraction of our careers.

The Regulation of Recombinant DNA Research

Perhaps because I am a woman or perhaps because I was close to Washington, but most certainly because I was one of the first plant molecular biologists, I was pulled into the recombinant DNA regulatory issues early in my career. My first adventure in Washington committees was serving on a scientific advisory panel on applied genetics convened by the then extant Office of Technology Assessment of the US Congress, asked to assess the future growth of US agricultural productivity. By the next year, I was a member of the new NIH Recombinant DNA Advisory Committee

(RAC), convened by the NIH Director to provide him with advice how recombinant DNA research should be carried out (Wivel 2014).

The RAC was organized by the NIH in response to concerns raised by several prominent scientists in letter to Science (Berg et al. 1974) and following a widely publicized conference convened at the Asilomar Conference Center to discuss laboratory containment conditions appropriate to the limited state of knowledge about the then new recombinant DNA (rDNA) technology. By the time I joined the RAC, it had elaborated and published a set of guidelines for the conduct of recombinant DNA research and had established itself as the go-to place for both academic and industry scientists to present and discuss not just proposed experiments, but the containment conditions under which they were to be carried out.

The RAC was a purely advisory committee to the NIH Director and had no ability to enforce its decisions. Nonetheless, just about everyone played by its rules, both companies and academic scientists. The few discovered violators lost NIH funding, but more importantly, lost the respect of their colleagues. That said, the initial guidelines were purely conjectural and based on the guesses of Asilomar conference participants and the initial RAC members about what might be dangerous. Perhaps the most critical aspect of how the committee functioned was its ability to consider new information, propose modifications to the guidelines, publish them for public comment and finally implement the modifications. Because rDNA technology proved so powerful in understanding genes, it was adopted extremely rapidly within the scientific community, hence experience with rDNA organisms accumulated rapidly. Because none of the conjectured problems emerged, the RAC was able within a few years to decrease the stringency of the containment conditions required for experimentation with rDNA and to progressively exclude many kinds of experimentation from the requirement for more than good laboratory practice.

As well, the RAC was able to re-examine some of the total prohibitions articulated in the earliest versions on the guidelines. For example, the initial guidelines forbade the cloning of toxin genes, such as those encoding the botulinus toxin, into *Escherichia coli* vectors. However, when the RAC convened a committee of researchers who worked the organisms that produce such toxins, it became apparent that cloning a toxin gene was, in fact, the safest way of working with it in the laboratory, because the laboratory strains of *E. coli* lack the genetic apparatus to colonize the gut and deliver the toxin to target tissues.

Throughout this early period, rDNA technology received quite a lot of publicity and generated considerable controversy. Perhaps the most outspoken critic of rDNA research in that period was Jeremy Rifkin, author of a 1977 book titled "Who Should Play God" (Howard and Rifkin 1977). Rifkin came to some of the RAC meetings, always making his views known. During my final year on the RAC, I was also a Phi Beta Kappa Visiting Scholar and spoke to both academic and community audiences about the substance of rDNA science, as well as the surrounding controversies, later publishing a written version of my talk in the Syracuse Scholar (Fedoroff 1986). During a visit to the University of Ohio, I met with faculty members engaged in rDNA research in animals, in addition to giving a radio interview and a public lecture. After my visit, letters from that part of the country began to come into the RAC expressing clear support of rDNA research, whose promise for medicine was already apparent—and would only grow in the future (I do not take credit for this, as I believe someone in the community organized the letter-writing campaign). The RAC's chairman was able to point to a stack of several hundred supportive letters in answer to Rifkin's next diatribe against rDNA research.

The first requests to field test transgenic plants began to come to the RAC in approximately 1983. The most striking result was that individuals from several regulatory agencies began to come to RAC meetings. As well, I began to be invited to attend meetings with regulators from the U. S. Department of Agriculture (USDA) and the Environmental Protection Agency (EPA). While the RAC continued to receive requests to approve research involving plants, it became evident that the EPA, the USDA, and to some extent the Food and Drug Administration (FDA) viewed transgenic plants as within their regulatory purview. By 1984, I had completed my term on the RAC and had become a member of both the Commission on Life Science and the Board on Basic Biology of the National Academy of Science (NAS), both of which had considerable interest in the regulation of rDNA research.

In due time, the Office of Science and Technology Policy convened a taskforce to coordinate the regulatory interests of the various agencies that had an interest in the use of rDNA technology in their domains, including the EPA, the USDA and the FDA. The committee produced a document titled the "Coordinated Framework for the Regulation of Biotechnology," which in 2016 remains the regulatory framework under which all three agencies continue to function (OSTP 1986). Tragically for biotechnology, particularly agricultural biotechnology, the actual regulatory practice evolved in an unintended direction.

In roughly the same timeframe, the Council of the NAS convened a small committee to produce a "white paper" expressing the NAS Council's position on the issues surrounding the introduction of rDNA organisms into the environment. I was asked to serve on the committee, chaired by plant pathologist and NAS Council member Arthur Kelman. The committee was small, but diverse, consisting of Wyatt Anderson, Stan Falkow, and Si Levin in addition to Kelman and me. After many rewrites and much discussion, we produced the promised white paper, titled "Introduction of Recombinant DNA-Engineered Organisms into the Environment: Key Issues" (Kelman et al. 1987).

The main conclusions of the white paper were as follows. First, there was then (and still is) no evidence that unique hazards attend the use of rDNA techniques or the movement of genes between unrelated organisms. Second, that the risks associated with the introduction of genetically engineered (GE) organisms into the environment were the same as those attending the introduction of unmodified organisms or organisms modified by more traditional techniques (which include chemical and radiation mutagenesis). And third, that assessment of risks of introducing GE organisms into the environment should be based on the nature of the organisms and the environment into which it was being introduced, not the method by which it was produced.

The OSTP committee basically subscribed to the same view on GE organisms as the NAS committee and therefore felt that no new legislation was necessary for organisms modified by rDNA techniques. The committee directed each of the three relevant agencies, the EPA, the USDA, and the FDA to find existing legislation under which to oversee the introduction of GE organisms into the environment and to make it product-, and not process-based. What actually happened is that only GE organisms were subjected to a high level of regulatory scrutiny, some spending years in regulatory purgatory before being approved. Moreover, EPA regulated under laws written to regulate toxic substances and the USDA regulated them as if they were plant pathogens. Only the FDA formulated a policy that was based on composition and not on process—and may not be able to stick to that decision, as evidenced by the recent congressional directive seeking to compel the FDA to mandate labeling of genetically modified salmon, despite the FDA's clear policy that labeling is reserved for ingredients with health or environmental implications.

The GMO Wars

Following the publication of the NAS white paper, I wrote a number of editorials about GE organisms for the NAS, as well as a New York Time Op Ed piece titled "Impeding genetic engineering" that appeared in the 2 September issues of 1987. I was still naïve enough to believe that the controversies would die down as people understood the technology and began to benefit from it. I took every opportunity offered to speak and write about these new molecular techniques of genetic modification. In 2001, I began discussing writing a book about genetically modified organisms, aka GMOs, with Nancy Marie Brown, a wonderful science writer then working for Penn State. It was to be focused on food and intended for a general audience. In due time, we became a good writing team: I did the research and provided the science, while Nancy deconvoluted my sentences and wrote readable stories around the science. The book, titled "Mendel in the Kitchen: A Scientist's View of Genetically Modified Foods," was published by the Joseph Henry Press of the National Academy of Sciences in 2004.

The publication of my book thrust me deeper into the controversies, already quite fierce, which have continued to surround the use of molecular modification techniques in plants and animals destined for food, that is, agricultural biotechnology. Many, although far from all, new technologies generate some resistance and some scare stories. So, for example, cell phones were rumored to produce brain tumors and food irradiation was suspected to make foods radioactive. With time and experience, many of these early concerns die down—who even remembers the cell-phone rumors when chasing the latest Apple iPhone? This is true even in the genetic engineering realm: what would we do today without the many drugs—

perhaps most prominently recombinant human insulin—produced using rDNA technology?

I first encountered the intensity of people's feelings about GMOs when I was touring the country promoting my book in 2004. At one "science café" in Berkeley, California, an audience member rushed from the back of the room, grabbed the microphone away from me, and began to shout about how this new science was ruining the flour he used in his bakery (there was and is no GM wheat yet). The audience surrounded me, fearing that he was about to attack me. The occasional organic farmer showed up at other venues, often speaking passionately about the evils of GMOs. But what I most often encountered was people uncomfortable with new genetic "technology" in the food space and seeking to blame it for what they saw as the eroding taste of fruits and vegetables (that, of course, resulted from conventional breeding for increased yield and decreased perishability on the way to market over taste).

In hindsight, the discourse about GMOs was still relatively civil back then. The escalation of GMO discussions in the public domain from discomfort and suspicion about a new technology to a full-scale war of words happened rather gradually though the last decade of the 20th century and continues unabated in the second decade of the 21st. This happened, paradoxically, even as the GM crops were commercialized and adopted at record speeds in much of the world with substantial economic and environmental benefits and with no evidence of harm to people, animals or the environment. The reasons for the continuing chasm between how the public views GMOs and the scientific evidence on their safety and efficacy are both interesting and deeply disturbing.

First, the scientific and economic evidence on GM crops today (as of this writing, there are no GM animals in commercial agriculture or aquaculture). GM crops have been adopted at unprecedented rates since their commercial introduction in 1996. In 2014, GM crops were grown in 28 countries on 181.5 million hectares (James 2015). More than 90% of the 18 million farmers growing biotech crops today are smallholder, resource poor farmers. Farmers aren't fools: they migrate to GM crops because their yields increase and their costs decrease. The simple reasons that farmers migrate to GM crops are that their yields increase and their costs decrease. A recent meta-analysis of 147 crop studies conducted over a period of 20 years concluded that the use of GM crops had reduced pesticide use by 37%, increased crop yields by 22% and increase farmers' profits by 68% (Klümper and Qaim 2014).

The GM crop base remains very narrow, with the vast majority of GM acres devoted to corn, soybeans, cotton and canola, largely because these are the kinds of extremely widely grown commodity crops that can support the cost of both their development and the huge regulatory costs that attend deregulation. Until last year, when President Obama issued a directive to the regulatory agencies to reexamine their approach to the regulation of biotechnology, the regulation of GM crops and animals was still operating under the 1986 Coordinated Framework for the

Regulation of Biotechnology (OSTP 1986) on a case-by-case basis and applied only to crops modified by molecular techniques.

The overwhelming evidence is that the GM foods now on the market are as safe, or safer, than non-GM foods (Richroch 2013). There is still no evidence that the use of GM techniques to modify organisms is associated with unique hazards. The EU has conducted biosafety research for more than a quarter of a century, expending upward of €300 million on 130 research projects involving more than 500 independent research groups. A recent report on this research concluded simply that biotechnological (e.g. GM) approaches are no more risky than older methods of modification (http://ec.europa.eu/research/biosociety/pdf/a_decade_of_eu-funded_gmo_research.pdf). Every credible scientific body that has examined the evidence has come to the same conclusion (http://gmopundit.blogspot.com/p/450-published-safety-assessments.html).

To date, the only unexpected effects of GM crops have been beneficial. Many grains and nuts, including corn, are commonly contaminated by mycotoxins, which are toxic and carcinogenic compounds made by fungi that follow boring insects into the plants. Bt corn, however, shows as much as a 90% reduction in mycotoxin levels because the fungi that follow the boring insects into the plants cannot get into the Bt plants (Munkvold 2003). Interestingly, planting Bt crops appears to reduce insect pressure in non-GM crops growing nearby. Indeed, the widespread adoption of Bt corn in the U.S. Midwest has resulted in an area-wide suppression of the European corn borer (Hutchison et al. 2010).

While there is a global consensus among scientists and scientific organizations that GM technology is safe, the political systems of Japan and most European and African countries remain opposed to growing GM crops. Many countries lack GM regulatory systems or have regulations that prohibit growing and, in some countries, importing GM food and feed. In Europe, the regulatory framework is practically nonfunctional; only one GM crop is currently being grown and only two others have gained approval since 1990 when the EU first adopted a regulatory system (Kershen 2014).

But even countries such as the U.S. that have a GM regulatory framework, the process is complex, slow and expensive. U.S. developers must often obtain the approval of three different agencies, the EPA, the USDA, and the FDA, to introduce a new GM crop into the food supply. Bringing a GM crop to market, including complying with the regulatory requirements, was estimated to cost $135 million in 2011 (McDougall 2011). By contrast, crops modified by older techniques, including radiation and chemical mutagenesis require no regulatory oversight.

It could be argued that when the molecular modification technology was first introduced more than 30 years ago, a cautious approach that even exaggerated the potential for harm was justifiable, but it is increasingly clear that the persistence of such a highly precautionary approach is today politically driven (Smyth and Phillips

2014). Let's "follow the money." NGOs, most vocally Greenpeace and Friends of the Earth, have conducted vigorous and successful campaigns of misinformation about GMOs first in Europe, then around the world (Paarlberg 2014; Wesseler and Zilberman 2014). Patrick Moore, one of the co-founders of Greenpeace has pointed out that: "Greenpeace is clearly a big-money operation these days, as intent at feeding itself as any corporation is" (http://www.science20.com/science_20/cofounder_of_greenpeace_greenpeace_is_wrong_about_golden_rice-122754). Apel has written extensively on the economics of opposing GMOs, concluding: "The key players encompassed by the definition of 'opponent' of engineered crops reap billions annually from restricting agricultural biotechnology or the food that results. Indeed, more money can be made from restricting agricultural biotechnology than by delivering it" (Apel 2010).

Perhaps the most counterproductive—even sinister—development is the increasing vilification of GM foods as a marketing tool by the organic food industry (Schroeder 2014). The organic agriculture finds its roots in rural India where Sir Albert Howard, known as the father of organic agriculture, developed composting methods that could kill the pathogens that abound in human and animal waste (Fedoroff and Brown 2004). The organic movement grew in the UK and Europe during the early 20th century and was later championed in the US by Jerome Rodale, even as synthetic fertilizers use was increasing worldwide. With the establishment of organic retailers, such as Whole Foods and Wild Oats, the organic food business grew rapidly and certification organizations proliferated. In the 1990s, Congress established the National Organic Standards Board (NOSB) under the USDA through the Organic Food Production Act and charged it with developing national standards, which were eventually published in 2000 and often called the "Organic Rule" (http://www.ams.usda.gov/grades-standards/organic-standards). The Organic Rule expressly forbids the use of GM crops, antibiotics, and synthetic nitrogen fertilizers in crop production and animal husbandry, as well as food additives and ionizing radiation in food processing.

Organic food is food produced in compliance with the Organic Rule; the USDA's Organic Seal is a marketing tool that makes no claims about food safety or nutritional quality. But the organic food industry has systematically used false and misleading marketing about the health benefits and relative safety of organic foods compared with what are now called "conventionally grown" foods (Schroeder 2014). Indeed, organic marketers represent conventionally grown foods as swimming in pesticide residues, GM foods as dangerous, and the biotechnology companies that produce GM seeds as evil, while portraying organically grown foods as both safer and more healthful. Recent "labeling" campaigns have the objective of promoting the organic food industry by conveying the message to consumers that food containing GM ingredients is dangerous. Mainstream media frequently carry messages that are positive about organic food and extremely negative about GM foods through programs like the Dr. Oz Show and through popular food writers such as Michael Pollan and Mark Bittman (Schroeder 2014). Simply said then, there's money and fame in being anti-GMO.

Why I Continue to Talk About GMOs

Over the almost four decades since my chance encounter with Barbara McClintock at Cold Spring Harbor focused my early career on inventing and using molecular techniques in plant biology—now, of course, called plant biotechnology—I have seen plant biologists amass vastly more knowledge about how plant genetic and physiological systems work at the biochemical and molecular levels than in all of the previous history of botany. I have appreciated contributing at many different levels, from developing techniques, to starting the whole field of plant transposon-ology, particularly transposon tagging, to explaining GMOs to anyone who would listen. In 2007, perhaps as a result of my continuing service on various Washington committees, I received an invitation from the President of the National Academy of Sciences to be a candidate for the position of Science and Technology Adviser to the Secretary of State, then Condoleeza Rice. Thinking this might give me a bit of an international bully pulpit from which to talk about GMOs, I agreed and, somewhat to my surprise, got the job. A few weeks into that new job, I was invited by Henrietta Fore, then the Administrator of the U.S. Agency for International Development (USAID) to be her science adviser, as well.

Wherever I traveled, with the noted exception of France, my scientific expertise on biotechnology was welcomed. I traveled to many countries, spoke to many audiences, and did endless press interviews on science, science diplomacy—and most of all, why GMOs and biotechnology matter to the world. My own word view expanded suddenly in 2009 when I attended a small, very international meeting on the impact of climate change on agriculture in Spitzbergen, Norway. Below is what I wrote for Andy Revkin's Dot Earth blog on the airplane coming home:

Dear Andy,   27 February 2009
I write you from the far frozen north of Norway. Near the village of Longyearbyen, on the island of Spitzbergen, is a remarkable structure called the Svalbard Global Seed Vault. In chambers deep in a mountain whose temperature never rises above freezing is a storage chamber, further cooled to a temperature of $-15\ °C$. In it are seeds of some 70,000 varieties of 64 of the world's major food crops. Marking the Vault's first anniversary, a small scientific meeting focused on how climate change will affect humanity's ability to grow food.

Spitzbergen

Longyearbyen

Svalbard Global Seed Vault

It's cold here. But a deeper chill settled on us as we listened to the climate scientists' scenarios for the coming decades. Even if we all stopped driving, flying and turned every light out tomorrow, the CO_2 we've already poured into our atmosphere over the last 100 years means that next hundred will be much hotter than the last 100.

It seems we've begun to absorb the notion of hotter, drier summers, rising sea levels and more extreme weather—bad enough. But who's thought much about what a changing climate might mean at the grocery store? No matter whose projections, no matter whether the best, worst or most probable scenario, our crops will suffer—and I mean OUR crops, not just those in some distant land.

Here's a real example of what a higher temperature can do. In 2003, France and Italy had a summer that was just 3.5% hotter than their usual summer. It rained as usual, but the yields of major crops were still down by 20–36%. Projections show that this will be the average summer by 2090.

And within the next few decades, it's nearly certain that we'll be recording summer hotter than ever recorded. Many of our crops fail completely if the temperature goes much above 100 °F for just a few days at a critical flowering time. During my days as a corn geneticist, I watched the tassels turn brown and sterile the summer it hit 108° in Columbia, Missouri at pollination time.

To put this in perspective: the food crisis of 2008 called attention to how close we are to the limits of the global food supply. But unlike the financial one, the food crisis isn't going away. This is because the number of people on the planet is still growing and by midcentury we'll need to roughly double the food supply—which, of course, starts with growing crops, whether to feed us or to feed to pigs and cows and chickens. Yet the amount of land on the planet that's good for growing crops hasn't changed much for more than half a century.

Will the warming climate open lands for cropping farther north? Probably, though how much is uncertain. What is quite certain by now is that climate change will squeeze those farther south as soil moisture declines. This will affect the most populous countries, countries whose populations are growing fastest.

So what do we do? The Global Crop Diversity Trust (http://www. croptrust.org/main/arctic.php), which funds the Vault, is dedicated to the preservation of the genetic diversity in our food crops. They are motivated by the belief that it is this diversity that will be the source of the genes we will need to develop plants that can grow on a hotter, drier planet.

Maybe. But the fact is that over the entire (more than) 10,000 year history of agriculture, the CO_2 levels in the atmosphere were between 180 and 280 ppm. We're at 389 now. It's not unlikely that we'll hit 700 ppm before we get this problem under control. That's going to mean that it'll get hotter and drier than anything but desert plants have seen before. Desert agriculture isn't new, but scaling it up will be a challenge and may mean venturing outside the limits of our current stable of crops.

I rather think we should also be scrambling to explore and understand organisms—not just plants—that have evolved to survive and thrive in the parts of the earth that are already the hottest and driest. We'll need to understand how they survive. We'll need to capture the genes that make it possible. If we're lucky, we'll be able to use these to arm some of our super-productive crops plants to survive and thrive under such conditions. If we're lucky. And if people stop being so reluctant to use modern molecular science to enhance crops.

Cheers! Nina

Upon my return to Washington, I organized an interagency workshop on the impact of climate change on agriculture, a subject that still receives little attention, even though severe weather is making us increasingly aware that the climate is changing. The speakers ranged from climate scientists to water experts to biotechnologists to conventional agricultural experts. Together we penned a piece titled "Radically rethinking Agriculture for the 21st Century" for Science magazine (Fedoroff et al. 2010). We judged the overall challenge of feeding a still-growing human population, already numbering more than 7 billion, on a warming planet as perhaps the greatest challenge of the century.

As I approached the end of my term at the State Department, I was invited to attend the opening of the King Abdullah University of Science and Technology (KAUST), as well as to present a paper in their opening science symposium. I had participated in the early phases of KAUST's organization just a few years earlier, before I entered the State Department and had to step away from that advisory role. I attended the spectacular opening on behalf of the State Department and presented a paper that synthesized concepts about important innovations in agriculture nec-essary to assure a food-secure future. Upon leaving the State Department, I was invited to join the KAUST faculty. I proposed to establish a Center for Desert Agriculture (https://cda.kaust.edu.sa/Pages/Home.aspx) and spent the next 4 years doing so.

I continue to talk and write about GMOs and the future of agriculture. In 2014, I was invited to give a TEDx talk (https://www.youtube.com/watch?v=fqJAeReFr8I)

at CERN in Geneva, giving me an opportunity to assemble an overview from an agricultural perspective of what falsified Thomas Malthus' gloomy prediction two centuries ago that humanity was doomed to famine and strife because population growth would always outstrip its ability to produce food (Malthus 1798). The answer, of course, is the rapid introduction of science and technology into agriculture in the 20th century (Fedoroff 2015). And yet in some sense, agriculture is threatened in the 21st century not just by population growth and climate change, but by its very success in freeing ever more people from the back-breaking labor of agriculture and vastly increasing its capacity to produce food.

The shift of population from rural to urban has been dramatic in the developed world, with less than 2% of the population today supplying the food for the rest. Most of us are urban today and have access to food through a vast global food system that supplies our food retailers with abundant produce. We are far removed from agriculture, blind to its basics and vulnerable to the increasingly strident opponents of modern agriculture who have successfully used fear to promote their own economic interests.

It is ironic that just as we have developed methods to make agriculture more sustainable and less reliant on the kinds of chemical inputs that have collateral damage, we have begun to turn on the very science that underpins our abundant, safe food supply. We certainly have the tools to vastly improve both the productivity and the sustainability of agriculture. GPS-guided, programmable farm machinery can markedly reduce water, chemical and energy use in agriculture. But today's affluent urban dwellers denigrate efficient, large-scale farming with the moniker "factory farming." We have the tools to modify crops to meet the challenges of climate change and shifting disease distributions, but the organic food marketers are doing their best to turn consumers against foods containing GM ingredients. And the regulators are still caught in the precautionary trap, unable to convert the evidence that foods with or from GM crops and animals are as safe or safer than any we've ever introduced into our food supply. Unfortunately, belief systems, particularly those around food, are very difficult to dislodge, however large the mountain of available facts.

References

Apel A (2010) The costly benefits of opposing agricultural biotechnology. N Biotechnol 27 (5):635–640. doi:10.1016/j.nbt.2010.05.006

Baker B, Schell J, Lorz H, Fedoroff NV (1986) Transposition of the maize controlling element *Activator* in tobacco. Proc Natl Acad Sci USA 83:4844–4848

Baker B, Coupland G, Fedoroff N, Starlinger P, Schell J (1987) Phenotypic assay for excision of the maize controlling element Ac in tobacco. EMBO J 6:1547–1554

Berg P, Baltimore D, Boyer HW, Cohen SN, Davis RW, Hogness DS, Nathans D, Roblin R, Watson JD, Weissman S, Zinder ND (1974) Letter: potential biohazards of recombinant DNA molecules. Science 185(4148):303

Fedoroff N (1983a) Notes on cloning maize DNA. Maize Genet Coop Newslet 57:154–155

Fedoroff N (1986) The recombinant DNA controversy: a contemporary cautionary tale. Syracuse Scholar 7:19–33

Fedoroff N, Wessler S, Shure M (1983) Isolation of the transposable maize controlling elements *Ac* and *Ds*. Cell 35:243–251

Fedoroff N, Furtek D, Nelson O (1984) Cloning of the *Bronze* locus in maize by a simple and generalizable procedure using the transposable controlling element *Ac*. Proc Natl Acad Sci USA 81:3825–3829

Fedoroff NV (1983b) Controlling elements in maize. In: Shapiro J (ed) Mobile genetic elements. Academic Press, New York, pp 1–63

Fedoroff NV (2015) Food in a future of 10 billion. Agricul Food Secur 4:1. doi:10.1186/s40066-015-0031-7

Fedoroff NV, Brown DD (1978) The nucleotide sequence of oocyte 5S DNA in Xenopus laevis. I. The AT-rich spacer. Cell 13:701–716

Fedoroff NV, Brown NM (2004) Mendel in the kitchen: a scientist's view of genetically modified food. Joseph Henry Press, Washington, DC

Fedoroff NV, Battisti DS, Beachy RN, Cooper PJ, Fischhoff DA, Hodges CN, Knauf VC, Lobell D, Mazur BJ, Molden D, Reynolds MP, Ronald PC, Rosegrant MW, Sanchez PA, Vonshak A, Zhu JK (2010) Radically rethinking agriculture for the 21st century. Science 327 (5967):833–834

Howard T, Rifkin J (1977) Who should play God? The artificial creation of life and what it means for the future of the human race. Delacorte Press, New York

Hutchison W, Burkness E, Mitchell P, Moon R, Leslie T, Fleischer SJ, Abrahamson M, Hamilton K, Steffey KL, Gray M (2010) Areawide suppression of European corn borer with Bt maize reaps savings to non-Bt maize growers. Science 330(6001):222–225

James C (2015) Global status of commercialized biotech/GM crops: 2014. ISAAA, Ithaca, NY

Kelman A, Anderson W, Falkow S, Fedoroff N, Levin S (1987) Introduction of recombinant DNA-engineered organisms into the environment: key issues. National Academy of Sciences, Washington, DC

Kershen DL (2014) European decisions about the "Whack-a-mole" game. GM Crops & Food 5 (1):4–7

Klümper W, Qaim M (2014) A meta-analysis of the impacts of genetically modified crops. PLoS ONE 9(11):e111629

Levy AA, Meissner R, Chague V, Zhu QH, Emmanuel E, Elkind Y (2000) A high throughput system for transposon tagging and promoter trapping in tomato. Plant J 22(3):265–274

Malthus T (1798) An essay on the principle of population. J. Johnson, St. Paul's Church-Yard, London

McClintock B (1987) The discovery and characterization of transposable elements. Garland Publishing Inc., New York

McDougall P (2011) The cost and time involved in the discovery, development and authorisation of a new plant biotechnology derived trait. Crop Life Ints

Miller JR, Cartwright EM, Brownlee GG, Fedoroff NV, Brown DD (1978) The nucleotide sequence of oocyte 5S DNA in Xenopus laevis. II. The GC-rich region. Cell 13:717–725

Munkvold GP (2003) Cultural and genetic approaches to managing mycotoxins in maize. Annu Rev Phytopathol 41(1):99–116

OSTP (1986) Coordinated framework for regulation of biotechnology. Federal Register 51:23, 302–323, 393

Paarlberg R (2014) Consequences of the anti-GMO campaigns. Paper presented at the Bread and Brain, Education and Poverty, Vatican, 4–6 Nov 2013

Parinov S, Emelyanov A, Gao Y, Naqvi NI (2006) Trans-kingdom transposition of the maize dissociation element. Genetics 174(3):1095–1104. doi:10.1534/genetics.106.061184

Qu S, Jeon JS, Ouwerkerk PB, Bellizzi M, Leach J, Ronald P, Wang GL (2009) Construction and application of efficient Ac-Ds transposon tagging vectors in rice. J Integr Plant Biol 51 (11):982–992. doi:10.1111/j.1744-7909.2009.00870.x

Richroch AE (2013) Assessment of GE food safety using '-omics' techniques and long-term animal feeding studies. New Biotechnol 30:351–354

Schroeder J (2014) Organic marketing report. Academic Rev

Shure M, Wessler S, Fedoroff N (1983) Molecular identification and isolation of the Waxy locus in maize. Cell 35:235–242

Smyth SJ, Phillips PW (2014) Risk, regulation and biotechnology: the case of GM crops. GM Crops Food 5(3):170–177

Van Sluys MA, Tempe J, Fedoroff N (1987) Transposition of the maize *Activator* element in *Arabidopsis thaliana* and *Daucus carota*. EMBO J 13:3881–3889

Weil CF, Kunze R (2000) Transposition of maize Ac/Ds transposable elements in the yeast saccharomyces cerevisiae. Nat Genet 26(2):187–190

Wesseler J, Zilberman D (2014) The economic power of the Golden Rice opposition. Env Develop Economic 19(06):724–742

Wivel NA (2014) Historical perspectives pertaining to the NIH recombinant DNA advisory committee. Hum Gene Ther 25(1):19–24. doi:10.1089/hum.2013.2524

Agriculture in the Modern Age
Biopesticides and Plant Biotechnology

Sue MacIntosh

Plant Biotechnology

My first work in plant biotechnology was in 1987, where I applied my protein biochemistry skills for the *Bt* group at Monsanto. I had spent the prior dozen years learning all about protein biochemistry, by purifying blood coagulation proteins, first at the University of Iowa (my alma mater) in Whyte Owens' lab, then at Sigma Chemical, where I shared responsibility for purification of the entire line of blood coagulation proteins found in the Sigma catalog. I jumped at the opportunity to work at Monsanto, leaving behind blood proteins for Bt proteins, a very interesting set of proteins.

David Fischoff led the molecular biology group, working to introduce *Bacillus thuringiensis* proteins into tomato plants, used as a model system. I worked in Roy Fuchs' lab, the analytical team that characterized what Dave's team had accomplished. I joked with the molecular biologists that the action was always on the side of protein biochemistry—because after all, proteins are the workhorses, the genes just produce the proteins.

First a bit more about Bt… Bt proteins are naturally found in the gram-positive, spore-forming *Bacillus thuringiensis* (Bt) bacterial strains, which formed large parasporal crystals thanks to the very high level of expression in the bacterium (Fig. 1). The activity of the Bt strains is not only due to the Bt proteins, but also to spores produced during the fermentation, in addition to soluble proteins and factors, such as VIP, zwittermycin, and others. Bt organisms were discovered in the early 1900s, and found to have excellent bioactivity against many insect pests. One of the very first commercial products was DiPel™ based on the *Bt kurstaki* strain, which demonstrated selective activity against a range of Lepidopterans but only at the larval stage. DiPel was named by reversing the first letters of its target pest: **LEPID**

S. MacIntosh (✉)
MacIntosh & Associates Inc., 1203 Hartford Avenue, 55116-1622 Saint Paul, MN, USA
e-mail: suemacintosh@mac.com

© Springer International Publishing AG 2017
L.S. Privalle (ed.), *Women in Sustainable Agriculture and Food Biotechnology*,
Women in Engineering and Science, DOI 10.1007/978-3-319-52201-2_4

Fig. 1 Bt kurstaki crystals by
scanning electron microscopy

opteran. Since that time, a number of different Bt strains have been marketed that
allow control larvae of lepidopterans, coleopterans and mosquitos. Today, these Bt
strains remain the cornerstone of insect control in organic production.

Amazingly, it has been estimated one third of the bacterial mass of a Bt strain is
made up of these Bt crystal protein(s). In the Btk strain, the crystals are bipyrimidal
in shape, remarkably composed of several different Bt proteins Cry1Aa, Cry1Ab,
Cry1Ac and Cry2. The Cry1 proteins are very unique, large protoxins with a mass
of >120 kDa and only soluble in very high pH (>10). They become active via
trypsin cleavage to roughly half of the original mass, occurring in the high pH of the
lepidopteran gut, binding to receptors and poking holes in the insect gut that ultimately kills the insect (MacIntosh et al. 1991). Today more than 600 different Bt
proteins have been isolated (http://www.btnomenclature.info/), but only a handful
has found their way into plants for market introduction. Bt proteins were an obvious
choice for biotech crops, given the long history of safe use and ease in cloning.
While Bt can be effective in spray formulations, the application had to be timed
precisely, because once the insect chewed into the crop (e.g., European corn borer,
stem borer, etc.), it would be protected from such sprays. Also, larvae were most
susceptible when they were young, at early instars.

There were a number of hurdles in applying Bt technology to crop plants. First,
was to decide which Bt protein is the best candidate for a given crop. Insect

bioassays using purified Bt proteins help solve this dilemma (MacIntosh et al. 1990b) by defining the insect spectrum of each protein, informing the best crop/Bt combination. Next, the expression of Bt proteins was very low in plants and activity could only be tracked using insect bioassays, sensitive silver staining and/or western blots. Introducing the large protoxin of >120 kDa undoubtedly hampered expression, and work continued using the much smaller tryptic fragment, yet expression levels were still below the level needed for adequate insect control. A number of different efforts were launched to enhance the activity of these Bt proteins including potentiators such as enzyme inhibitors (MacIntosh et al. 1990c) and modifying the genetic codes. Unfortunately, enzyme inhibitors have properties of anti-nutrients, which would create a rather large regulatory barrier.

One of the biggest scientific advances was the use of synthetic gene technology to boost protein expression. While one of the first applications of the synthetic gene technology was the Bt protein expression in plants, it now is a standard technique of modern biotechnology. Scientists realized that the genetic G:C ratio of most bacterial expressed proteins was vastly different from the genetic codes of plant expressed proteins. Once the G:C ratio of Bt genes was adjusted to better match that of the plant, while at the same time keeping the important amino acid sequence unchanged, the expression level jumped some 500-fold (Perlak et al. 1991). Bt plants were now providing commercial levels of insect protection. It was an exciting time.

Resistance Management

With the success of Bt crop transformation, came the risk of insect resistance. Bt crops were very effective in controlling their target pests, however, insects are also very good at evolving resistance to any control measure. The crop protection industry (i.e., synthetic pesticides, biochemicals and biopesticides) has been in an ongoing battle to stay ahead of insect resistance. The industry has active discovery programs in search of new classes of insecticides, and has developed insect management strategies that assist growers to manage insecticide use in order to preserve all insecticides. Despite these efforts there remains a long list of insects resistant to many different crop inputs.

A number of strategies have shown value, such as active ingredient rotation, pesticide mixtures and other mechanical methods. Resistance was documented for sprayable Btk formulations, so the potential for resistance evolution was real. I was involved in unlocking the mystery surrounding Bt resistance by understanding the underlying mechanisms of action of Bt proteins which led to development of robust resistance management strategies for Bt crops (MacIntosh et al. 1991; Marrone and MacIntosh 1991, 1993; Stone et al. 1991; MacIntosh 1998, 2009; Ferré and MacIntosh 2007; Ferré et al. 2008).

The EPA took a strong stance by insisting that all Bt crop developers establish a means to delay the onset of insect resistance prior to market introduction. This forced developers to work closely with government and academic researchers to establish a workable plan. This was my first experience with industrial politics and it was an eye-opener. Industry joined government and academic scientists to support research to understand insect biology of target pests, such as insect movement both at the larval and adult stages, reproductive efficiencies and the genetics of resistant insects. Deciding the best approach to resistance management caused tremendous disagreements within industry as many companies were against applying limits to Bt crops sales and claimed that growers could not manage complicated growing schemes. I was lucky to be in the 'other' group, from a company that wanted to align with the academic scientists and adopt the best scientific advice to prolong the usefulness of Bt crops (Caprio et al. 1999). Ultimately, resistance management plans were developed, although in the early years (late 1990s) there was a patchwork of different plans that did nothing but confuse growers. Finally, the plans were consolidated and made uniform across each Bt protein/crop combination.

In most cases, plants that expressed high doses of the Bt protein were coupled with refuges of non-Bt plants grown close to, or within a Bt crop field. When possible, the expression of 2 Bt proteins against the same major insect pest provided even greater chance of delaying resistance development. When refuges of non-Bt plants were deployed, they allowed for a susceptible population of insects to be established, which escaped selection pressure from the Bt plants). As long as sufficient numbers of susceptible insects were available to mate with the rare resistant insect, the risk of resistance would be reduced, prolonging the efficacy of Bt crops. EPA also required annual monitoring for insect susceptibility and grower compliance surveys. The first Bt maize crop was marketed in 1996. Twenty years later, these resistant management plans have proved to be successful in delaying resistance for most, but not all, deployed Bt crop plants.

Sprayable Bts, Organic and the BioImage Satellite Laboratory

Having survived the challenges, exciting discoveries and conflicts around Bt crops I also realized that working in a large multi-national company with male-centric management was likely limiting my career. I entered Monsanto with a B.S. in zoology/minor in chemistry. Without a Ph.D., upward movement was unlikely as demonstrated by a sub-par position, still below a starting Ph.D. position, despite being first author to four publications during my three years of research (MacIntosh et al. 1990a, b, c, 1991). While I was grateful for the exciting and productive years at Monsanto, I decided to continue with Bt research and accept an opportunity in the world of sprayable Bt formulations, moving my two daughters to the Davis CA area and Novo Nordisk Entotech (Entotech), headed by fellow Monsanto alum, Pam Marrone. She promised me opportunities, and working for a woman seemed to be a good option; I was not disappointed. Entotech was wholly owned by Novo Nordisk (Novo), a Danish company, primarily involved in pharmaceutics and enzymes (i.e., insulin, human growth hormone, etc.) and served as the discovery arm for their biological pesticide business based on Bt strains. In the protein biochemistry group, we identified the active soluble small molecule, zwittermycin that enhanced the activity of Btk towards a typically recalcitrant insect, *Spodoptera* species. Interestingly, simply modifying the downstream processing of Bt fermentations could boost the zwittermycin and activity against *Spodoptera* species.

During this time, Entotech was approached by Ole Thastrup (Novo, Denmark) to participate in an internal Novo competition to develop a new technology that would be useful across the entire company. Given my interest in basic research, I was allowed to work with Ole and his team to develop the proposal. The winner of the competition would be given roughly $1M to establish a new lab in Denmark to develop a new technology. It was an interesting approach to excite the large group of Novo scientists, which upper management felt were becoming complacent with the success of insulin, their block buster product. The technology put forward by Thastrup was to develop high throughput video imaging (HTPVI) using various luminescent probes that could detect a multitude of different elements (i.e., Ca, Mg, etc.) under a range of experimental conditions.

Within our team, one scientist developed the idea of monitoring fermentation by using biomarkers. At that time, fermentation was the main technology to produce the various pharmaceuticals marketed by Novo, but tank monitoring was still rather simplistic and often a tank was lost because physiological changes could not be quickly identified and rectified. Another scientist sought to develop HTPVI as a research tool to better understand the mechanism of action of insulin, and the lack of sugar metabolism of diabetic patients. My project was to develop HTPVI screening methods to search for new insecticides, focusing on pore-formers, like that of Bt proteins. Thastrup's team met several times to develop our thoughts and plans around our proposed projects, building story boards with potential schemes for developing the HTPVI technology around these areas of interest. While in Denmark for a 6-week project to purify large quantities of zwittermycin, I participated actively in the discussions and plans for this competition. During the initial presentation in May 1993, teams presented their ideas and our HTPVI team was chosen to move to the next round. Our research plans were further refined with the ultimate presentation in front of the top 100 scientific managers in Novo, who would vote for their favorite research project.

My 6 weeks in Denmark concluded just a week prior to the final presentation and voting by top management. While I wanted to stay on, as a single parent, I needed to return home to my daughters. During the last meeting before departure, I sat with the Danes as we discussed our chance for success. I was a bit amazed at their negative attitude, especially since I am such an optimist, but they insisted that this other Novo team would win because of .... Or this other team would surely win, etc. I returned home certain that we had no chance. The following week, I received a very excited call from Dr Marrone, "You won, you won, you won....

You are moving to Denmark!" To which I responded, "I can't move across the world with two teenage daughters! No way!"

But, in about 2 weeks time, my mind was changed as I realized the opportunity was not one I could decline. As expected, my daughters were not thrilled about another move, but we started the process and took a trip to Copenhagen to find housing and schools. I planned the trip to do my best to convince them, seeing the little mermaid at the harbor, choosing a great house, the American-based school and visiting Tivoli, a beautiful park in the center of Copenhagen filled with rides, restaurants and flowers galore. We moved to Copenhagen October 1993 on a 5-year contract, renewable year by year.

Our time in Denmark was an incredible learning experience for us. The girls did well in the American-based schools and instantly grasped the train system, so important for getting to school and around Copenhagen. Working in Denmark greatly informed my international experience, especially the way in which work was conducted in a socialistic society. The lab was located in Søborg, away from the original Novo campus in Bagsværd, Denmark. Decisions about how the labs/offices would be arranged and outfitted included everyone—from the lab assistants to Thalstrup—everyone was heard and we met regularly until all the decisions were agreed and the lab was occupied. Because of this close cooperation, everyone was pleased with the new lab. Similar approaches were used for all major decisions, with the full team, allowing all views to be heard, discussed and evaluated. In this way, the team progressed in a unified and cohesive manner.

The insect project focused on establishing a cell-based assay system using Sf9 cells. Although Sf9 cells originated from *Spodoptera frugiperda* (fall armyworm), they were isolated from pupal ovary cells and not from the insect's gut. Yet, this cell line is susceptible to gut-based insect toxins such as Bt proteins, baculoviruses and other *S. frugiperda* insect toxins. The cell assay was established using a luminescent probe to visualize intracellular function. The intact Sf9 cells were loaded with the luminescent probe for calcium, washed, and bathed in a rich calcium environment creating a large gradient of calcium across the cell membrane, between the intact cell and extracellular solution. When Bt was added, pore formation was detected by an intense influx of calcium into the Sf9 cells, shown by a spectral shift of the luminescent probe (MacIntosh et al. 1994). This Sf9-based method could allow discovery screening for other pore-former insect toxins. However, after our first year, Novo decided to sell their biopesticide business and my time at the BioImage Satellite ended. I returned to Davis to assist in the sales of Entotech, leading to a large shift in my career.

The Regulation of Biotechnology-Based Crops

The end of the 1980s into the early 1990s saw an explosion of research on genetically modified plants. While I was engaged in Monsanto research, Calgene, CIBA-Geigy (now Syngenta), Plant Genetic Systems (PGS; Ghent, Belgium) and others were developing the first modify crops. By the mid-1990s, the first crops were being approved by regulatory authorities: Herbicide-tolerant tobacco in the EU, Bt potato by US EPA, herbicide-tolerant soybeans by USDA, etc. Associated with these actions was an increasing need for regulatory affairs staff with a strong scientific background that understood the technology and could assemble the scientific data to establish crop and protein safety.

In 1995, after the biopesticide business of Novo was sold, I joined PGS as their US Regulatory Affairs Manager working on GM corn products, including a molecular hybrid system and insect control. Over the next 5 years, PGS would be purchased by AgrEvo GmbH and then merged with Rhône-Poulenc to form Aventis CropScience in 1999. Finally, in 2002 Bayer purchased Aventis CropScience. Working within these 4 companies over 10 years was quite a challenge and taught me skills of focus and patience. During these years, I gained experience in nearly all possible regulatory affairs and regulatory science positions.

Returning to 1995, the Bt corn being developed at PGS expressed a new unique Bt protein, Cry9C, which was ultimately branded as StarLink™ Corn. Cry9C corn offered a new IRM tool for corn production since it was different from any of the Cry1 Bt proteins, binding to different receptors in the target insect gut cells (Fig. 2). Expression of Cry9C was robust using a truncated trypsin-resistant gene and the crop was agronomically and nutritionally equivalent to conventional corn. However, during the development project, it was discovered that Cry9C was stable to simulated gastric fluid (SGF) studies and therefore might have allergenic properties.

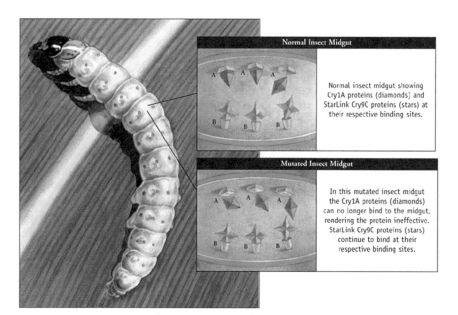

Fig. 2 Schematic of receptor binding of Cry1A and Cry9C proteins

The risk assessment for protein safety includes a wide range of different studies that not only characterizes the newly introduced protein, but also seeks to compare the protein with known toxins and allergens using bioinformatics (Delaney et al. 2008). The allergenicity assessment takes a multipronged approach since it is understood that no single test can predict if a substance will cause an allergenic response in sensitive people (Codex 2003). The most common food allergies are linked to 8 foods (e.g., milk, soybean, egg, wheat, tree nuts, peanuts and shellfish), with some of these foods expressing a large number of different allergenic proteins (e.g., soybean has 37 allergenic proteins). When a protein is sourced from one of these 8 food groups, additional studies ensure that this protein is not one of the allergens (Thomas et al. 2009). Allergenic proteins are typically found in rather high levels within the food, >1%, and are often stable. But not all stable proteins are allergens. With the availability of extensive protein databases, the introduced protein could be compared to allergenic proteins and standards were established that defined the acceptable level of similarity to guide the risk assessment (Thomas et al. 2005). Protocols for stability studies with simulated gastric/intestinal fluids and heat were also developed (Thomas et al. 2004).

The Cry9C protein showed no sequence similarity to toxins or allergens, was not expressed at high levels, but the stability to SGF raised concerns. I was responsible for the regulatory data package for GM corn projects, serving as Global Regulatory Affairs Manager and gaining approvals in all relevant countries. The authorities were excited about the IRM advantages StarLink Corn brought to the grower, and all three US authorities, EPA, FDA and USDA, decided to approve the product to limited acreage starting in 1998, with a restriction to animal feed use only—no food use allowed. During this time PGS (and successor companies) worked with allergy experts to run additional tests in an attempt to establish the safety of Cry9C and gain food approval. But before a final decision could be made for food approval, an anti-GM group found StarLink Corn in taco shells, and the product registration was voluntarily withdrawn and a recall for taco shells and StarLink grain began in Oct 2000. This was the first recall of a GM crop. Aventis initiated an extensive grain testing and buy-back program, several people lost their jobs and a number of lawsuits were filed. All suits were settled out of court and estimations of the cost to the company have been wide-ranging.

The year after the taco recall was arduous as an individual, within Aventis and across the fledgling biotech industry. There were however, several positive outcomes from this incident; (1) regulatory authorities no longer grant split approvals for food or feed, (2) communication channels opened between the biotech developers and food companies, (3) governments and industry gained a greater appreciation of the adventitious presence (AP) issue leading the Japanese authorities to allow 0.5% AP of an unapproved biotech event as long as it was approved in at least one OECD country, and (4) standardized testing protocols were developed to improve the allergen risk assessment for GM crops under the umbrella of the International Life Sciences Institute (SGF; Thomas et al. 2004). At the end of 2000, I was invited to be the founding chair of the Protein Allergenicity Technical Committee (PATC; under ILSI) that uses a unique three-legged stool approach where academic, industry and government scientists work together to solve regulatory issues. Since 2000, the PATC has published 15 peer-reviewed articles mostly on improved and standardized methods, provided expertise that strengthened relationships with regulatory authorities around the world and participated in global organizations such as Codex to improve the allergenicity risk assessment of GM crops (Thomas et al. 2009; http://hesiglobal.org/protein-allergenicity-technical-committee-patc/).

After this challenging episode at Aventis, I was tapped to re-constitute the US regulatory team in 2001, and soon headed up the regulatory science lab as well. I grew into this management position over the next few years, overseeing the whole America's regulatory group—both regulatory affairs and regulatory science for North/South Americas. We rebuilt our relationship with the authorities and worked toward success with new product development efforts. Moving up that corporate ladder, I ultimately becoming the Vice President of Regulatory Affairs for the BioScience division (Bayer CropScience 2003–2005), which also brought the dread of corporate politics back into my career.

In 2005, I left Bayer and returned home to the Midwest to open my own business, MacIntosh & Associates, Inc., an independent consultancy. The firm focuses on biosafety compliance services for clients developing biological-based products for the agricultural market. Strategic support, technical expertise, and regulatory services are provided to advance products from initial research stages through development onto commercial release and post-marketing requirements with special attention to compliance with government regulations at all stages. My clients include small start-up companies to major multinational corporations that are developing organic biopesticides and/or GM crops.

Over the last 11 years, I have successfully managed a healthy balance of family and work with over 50 different client companies. MacIntosh & Associates has assembled regulatory data packages and obtained more than 20 EPA registrations on behalf of their clients, listed most of those products under the National Organic Program. Other efforts assist companies by organizing regulatory studies, writing up risk assessments and providing strategic advice in support of the de-regulation of GM crops. Products have ranged from animal feed additives, organic nematode control products, bio-insecticides and herbicides, to GM products that produce vaccines, insect control, herbicide tolerance, altered oil profiles and even plants that glow in the dark. It amazes me the extensive diversity of products and ideas that my clients are developing for the agricultural marketplace. It is just this diversity that allows growers to choose the best options for their farm and production goals allowing agriculture to flourish.

In this advanced era of agriculture, with rising global populations and the associated food requirements, the ability to grow crops in an effectual manner become ever more important. Managing crop production is done at the local level, with a kaleidoscope of different cultural, mechanical, biologic and chemical inputs and practices, some of which are described above. It is this mixture of different methods that is the basis for robust integrated pest management and agricultural sustainability.

In some countries, such as the USA, cropping monocultures have been embraced providing unprecedented and ever increasing yields of corn and soybeans across the Midwest. Yet, some criticize this cropping approach because of the challenges to pest control and the potential for environmental harm (e.g., reducing soil health). The high adoption of biotech crops in the Midwest US has led to reduced tillage and even no-till acreage that preserves the valuable topsoil and at the same time reduces fuel costs. Herbicide tolerant crops have provided easier weed management, although resistant weeds are becoming an issue in certain regions. Both biopesticides and biotech crops have led to better insect control with reduced costs and provide safer options to the grower. Biopesticides and Bt crops both preserve non-target organism populations limiting the need for harsher synthetic chemical treatments.

Growers in many countries may grow several crops on their small plots allowing crop rotation and unique management practices to control pests. Despite the specific local production practice(s), a variety of different management tools are absolutely required for success. The discovery and development of fertilizers, GM crops, biostimulants, biopestides and synthetic pesticides are all part of the agriculture sustainability puzzle, thoughtfully managed by growers.

References

Caprio M, Luttrell R, MacIntosh S, Moellenbeck D, Rice M, Siegfried B, Sachs E, Stein J, Van Duyn J, Witkowski J (1999) In: An evaluation of insect resistance management in Bt field corn: a science-based framework for risk assessment and risk management. ILSI Press, Washington D.C

Delaney B, Astwood JD, Cunny H, Eichen Conn R, Herouet-Guicheney C, MacIntosh S, Meyer LS, Privalle L, Gao Y, Mattsson J, Levine M (2008) Evaluation of protein safety in the context of agricultural biotechnology. Food & Chem Tox 46:S71–S97

Ferré J, MacIntosh S (2007) Insect species of importance to currently deployed Bt-crops that have developed resistance to *B. thuringiensis* toxins in the laboratory, 3rd ecological impact of genetically modified organisms meeting. Warsaw Poland, Abstract, 23–25 May 2007

Ferré J, Van Rie J, MacIntosh S (2008) GM crops and insect resistance management (IRM). In: Romeis J, Shelton AM, Kennedy GG (eds) Integration of insect-resistant GM crops within IPM programs. Springer, US

Marrone PG, MacIntosh SC (1991) Insect resistance to biotechnology products: an overview of research and possible management strategies. In: Denholm I, Devonshire A (eds) Resistance '91: achievements and developments in combating pesticide resistance. Elsevier Applied Science, London, pp 272–282

MacIntosh SC (1998) A science-based development of a resistance management strategy. In: Proceedings of the 5th international symposium: the biosafety results of field tests of genetically modified plants and microorganisms (Schiemann, J., Ed), Arno Brynda, Berlin, pp 218–224

MacIntosh SC (2009) Managing the risk of insect resistance to transgenic insect control traits: practical approaches in local environments. Pest Manag Sci 66:100–106

MacIntosh SC, McPherson SL, Perlak FJ, Marrone PG, Fuchs RL (1990a) Purification and characterization of *Bacillus thuringiensis* var. *tenebrionis* insecticidal proteins produced in *E. coli*. Biochem Biophys Res Comm 170:L665–672

MacIntosh SC, Stone TB, Sims SR, Hunst PL, Greenplate JT, Marrone PG, Perlak FJ, Fischhoff DA, Fuchs RL (1990b) Specificity and efficacy of purified *Bacillus thuringiensis* proteins against agronomically important insects. J Invert Path 56:258–266

MacIntosh SC, Kishore GM, Perlak FJ, Marrone PG, Stone TB, Sims SR, Fuchs RL (1990c) Potentiation of *Bacillus thuringiensis* insecticidal activity by serine protease inhibitors. J Agric Food Chem 38:1145–1152

MacIntosh SC, Stone TB, Jokerst RS, Fuchs RL (1991) Binding of *Bacillus thuringiensis* to a laboratory-selected line of *Heliothis virescens*. PNAS 88:8930–8933

MacIntosh SC, Svendsen N, Thastrup O (1994) Exploring Bt mode-of-action with high resolution video imaging. In: Proceedings of the VIth international colloquium on invertebrate pathology and microbial control and IInd international conference on Bacillus thuringiensis (Bergoin, M., Ed.). Achevé d'imprimer, Ganges, France, pp 223–224

Marrone PG, MacIntosh SC (1993) Resistance to Bacillus thuringiensis and resistance management. In: Entwistle PF, Cory JS, Bailey MJ, Higgs SR (eds) Bacillus thuringiensis, an enviornmental biopesticide, theory and practice. Wiley, New York, pp 221–235

Perlak FJ, Fuchs RL, Dean DA, McPherson SL, Fischhoff DA (1991) Modification of the coding sequence enhances plant expression of insect control protein genes. PNAS 88:3324–3328

Stone TB, Sims SR, MacIntosh SC, Fuchs RL, Marrone PG (1991) Insect resistance to Bacillus thuringiensis. In: Maramorsch K (ed) Biotechnology for biological control of pests and vectors. CRC Press, Boca Raton, FL, pp 53–66

Thomas K, Aalbers M, Bannon GA, Bartels M, Dearman RJ, Esdaile DJ, Fu TJ, Glatt CM, Hadfield N, Hatzos C, Hefle SL, Heylings JR, Goodman RE, Henry B, Herouet C, Holsapple M, Ladics GS, Landry TD, MacIntosh SC, Rice EA, Privalle LS, Steiner HY, Teshima R, van Ree R, Woolhiser M, Zawodny J (2004) A multi-laboratory evaluation of a common *In Vitro* pepsin digestion assay protocol used in assessing the safety of novel proteins. Reg Tox & Pharm 39:87–98

Thomas K, Bannon G, Hefle S, Herouet C, Holsapple M, Ladics G, MacIntosh S, Privalle L (2005) *In Silico* methods for evaluating human allergenicity to novel proteins: international bioinformatics workshop meeting report, 23–25 February 2005. Tox Sci 88:307–310

Thomas K, MacIntosh S, Bannon G, Herouet-Guicheney C, Holsapple M, Ladics G, Mclain S, Vieths S, Woolhiser M, Privalle L (2009) Health & environmental sciences institute (HESI)'s protein allergenicity technical committee (PATC) scientific advancement of novel protein allergenicity evaluation (2000–2007). Food & Chem Tox 47:1041–1050

The Story Behind the Approval of the First *Bt* Maize Product

Laura S. Privalle

The Set-up

When the agricultural biotechnology center of CIBA-GEIGY was first set up they were given 10 years to have a product. I did not know this when I joined about eight months into the big experiment. All the large agrichemical companies were getting into the biotech business, following the lead of the oil companies like Exxon and Mobile. This was the early 1980s. For me, this was a great opportunity to work on plant biochemistry in a world devoted to improving plants with real applications on the horizon (something less esoteric than academia and with goals that my family could actually understand). I joined CIBA-GEIGY as a plant protein bio-chemist after a brief post-doctoral fellow stint at Duke University in the bio-chemistry department in the medical center. Mary-Dell Chilton was the leader of the biotech unit. Getting her was a big coup and shortly after I joined, she was inducted into the National Academy of Sciences—a huge achievement for a sci-entist not even 50 years old and especially so for a woman at that time.

My first project at CIBA-GEIGY was on ethylene biosynthesis—the thought being that if we could control this "ripening" plant hormone, we could influence the stability of fruit (such as tomatoes) after they were harvested. This was also the basis of the first biotech product marketed—the Flavr Savr® tomato by Calgene. As it turned out all the companies were working on basically all the same product concepts because they were all starting with the same knowledge base. Nobody was really sure what these products would be worth (hey we were scientists—not business people) but we knew if we had the gene, it could be introduced into a plant and give it a new characteristic. Of course, what we didn't know was how hard it

L.S. Privalle (✉)
Research Triangle Park, 27705 North Carolina, USA
e-mail: laura.privalle@bayer.com

© Springer International Publishing AG 2017   71
L.S. Privalle (ed.), *Women in Sustainable Agriculture and Food Biotechnology*,
Women in Engineering and Science, DOI 10.1007/978-3-319-52201-2_5

would be to introduce genes into crops that growers might actually be interested in buying, i.e. develop a product with added value for which someone would be willing to pay more. Plus ten years seemed like a long time. We also did not foresee that public acceptance would become such an issue.

The Technical Challenges

Some of the technical hurdles that had to be overcome were (1) identifying the trait; (2) isolating the gene responsible for this trait; (3) introducing it into a plant cell, (4) getting the plant cell to regenerate into a *fertile* plant; (5) getting the plant to actually express the gene (produce the protein encoded by the gene); (6) find the plant event that was efficacious (actually expressed the gene in a manner that met product specifications); and (7) figuring out how to bring the product to market (Fig. 1). As a plant biochemist, my role came at the beginning of the process— identifying the trait and the protein that if expressed in the plant would give a useful phenotype and also at the end of the process; figuring out how to demonstrate the safety of the product so that it could be accepted globally and brought to market. After about eight years working on the former, I have spent the ensuing time focused on the latter. By 1992, the scientists at CIBA (Martie Wright and Karen

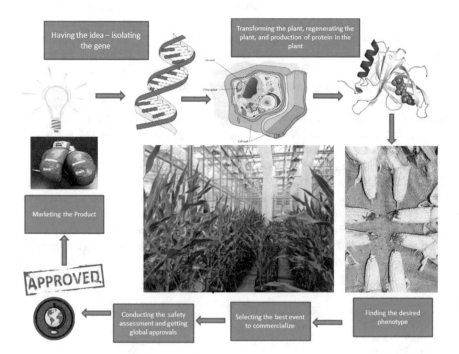

Fig. 1 Bringing a Biotech product to market

TRANSFORMATION USING BIOLISTICS

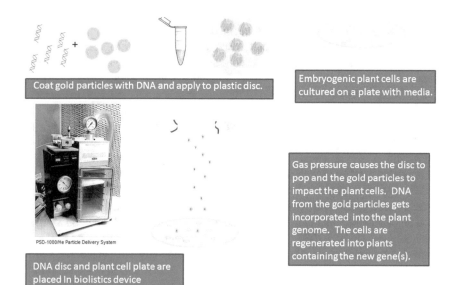

Coat gold particles with DNA and apply to plastic disc.

Embryogenic plant cells are cultured on a plate with media.

PSD-1000/He Particle Delivery System

Gas pressure causes the disc to pop and the gold particles to impact the plant cells. DNA from the gold particles gets incorporated into the plant genome. The cells are regenerated into plants containing the new gene(s).

DNA disc and plant cell plate are placed In biolistics device

Fig. 2 Transformation using Biolistics

Launis) had successfully introduced a gene (isolation of the gene, identification of the correct codons, the best expression elements, the best selectable marker, etc. involved contributions from many members of the team and while extremely critical will not be discussed here) into a corn cell using a biolistics approach (Fig. 2). The cell regenerated into a fertile plant and its progeny demonstrated resistance to the European corn borer (ECB, *Ostrinia nubulalis*). The race to get this approved and on the market was on!

ECB is a nasty pest of corn (maize), as the name suggests, it bores its way into the stalk of the plant which it then hollows out and the plant may eventually fall over (lodge) causing a negative impact on yield. The protein that the plant was to express to protect itself was from *Bacillus thuringiensis*. This is a so-called "*Bt*" protein, the product was *Bt* maize. Scientists had known about these very useful *Bt* proteins since the early 1900s when the *B. thuringiensis* was isolated from infected silkworms and the causal agent identified as this protein. The first U.S. registration of a *Bt* microbe as a commercial biological pesticide was in 1961 (Koch et al. 2015). By 1998 there were approximately 180 products registered in the United States and currently remain a favorite of organic farmers for control of Lepidopteran pests on many crops. The main problem was that when microbial products are applied, the sunlight tends to break down the *Bt* protein, so its effectiveness is short term. But, we now knew that if the plant itself made the protein, the control was much better. But how to demonstrate that this was a safe product?

The Safety Question

No biotech specific legislation was deemed necessary to regulate these kinds of products in the United States. Instead it was felt that sufficient legislation existed to cover these new kinds of products. There are actually five separate acts in place. Under FIFRA (Federal Insecticide, Fungicide, and Rodenticide Act), the Environmental Protection Agency (EPA) must approve all pesticidal products and set tolerance limits for exposure. *Bt* maize came under that category. The Plant Variety Protection Act (1970) empowers the Department of Agriculture (USDA) to regulate the planting, movement, import and export of biotech crops. So if the grower didn't want to apply for a permit to plant his field, the product needed to be "deregulated". NEPA (National Environmental Policy Act 1969) requires that the lead Federal agency conduct and publish an evaluation of the environmental impact of their action (such as approving a biotech product). The legislation that brings the Food and Drug Administration (FDA) into the picture is the Federal Food Drug and Cosmetics Act (FDCA 1938), which covers adulterants and additives, as well as the Toxic Substance Control Act. Biotech crops are generally not considered adulterants and are not toxic substances, so while all registrants seek a consultation with the FDA, the interaction is voluntary. Therefore, in the US, three regulatory agencies were involved. The specifics of what would constitute a robust safety assessment were undecided. However, *Bt* microbial products had already been determined to be safe, so safe that they were allowed to be used the same day that the product was harvested, so it was felt that there was a good basis on which to justify safety. We also thought that getting this first approval would involve many more studies than what would be needed as the world became more used to the idea of these kinds of products. We certainly did not foresee that more and more and more studies and parameters would be required.

It is important to keep in mind, that the class of chemistry involved here is the protein class. These first products involved the introduction of a gene, which encodes a protein which imparts the trait. For example, for herbicide tolerance, the new protein could be the target site of the herbicide but modified so that it is not sensitive to the herbicide. Alternatively, for the *Bt* crops, insect resistance is imparted by the expression of the *Bt* protein such that if the insect pest nibbles on the plant, the insect pest dies. This means that unlike generally broadcasting a pesticide where both beneficial insects, nontarget pests as well as target pests are all exposed—only those insects which actually consume the plant part expressing the protein would be impacted. A lot is known about proteins. All proteins, from plants, animals, microbes, whatever, are made up of the same 20 amino acids. Proteins are essential for humans. Humans have evolved a digestive tract that allows the consumption of the protein, its breakdown to its component amino acids and the absorption of those amino acids which the human then reassembles into its own proteins. The most common source of protein in our diet is meat, however, legumes and nuts are also good sources of protein. There are some proteins that are not good and are even harmful if consumed. Examples include botox (botulinin toxin), ricin, etc. However, a lot is known about these proteins so it is easy to avoid working with them.

The product we were working with was designed to express the *Bt* protein in green tissue and in the pollen as those were believed to be the first tissues that the young ECB larvae consumed as they encountered a maize plant. Also, we hoped that by not expressing the protein in the corn grain public acceptance would be facilitated. Transformation was accomplished using the "gene gun"; the so-called biolistics device that delivered DNA on gold particles using gas pressure acceleration (originally developed by Sanford et al. (1987) at Cornell University, it was modified by DuPont and sold by Biorad). This method was pretty brute force and tended to result in transformation events that had many fragments inserted into the genome. The one with the least complicated insertion site that retained efficacy was selected for commercialization. Unfortunately, one of the casualties of the delivery mechanism was that the promoter element driving the selectable marker was fragmented so that expression throughout the plant was quite low, below our lower limit of quantification.

So in 1992, we began our work on demonstrating the safety of our Bt maize product. We had access to information about one biotech product that had already been negotiating the approval process, Flavr Savr™ Tomato by Calgene. In this product, the introduced DNA encoded a selectable marker, APH3 and an antisense gene to prevent the expression of an endogenous gene involved in ripening. The "active ingredient" was actually not expressed. The registrant therefore focused on demonstrating the safety of RNA and of the selectable marker and sought approvals from the USDA and FDA. It was not a pesticidal product so was not reviewed by the EPA. However, they published the information that they presented to the regulatory agencies and this became a kind of bible for us. Of course, our product was different; it contained a selectable marker, the PAT (phosphinothricin acetyltransferase) protein encoded by the *bar* gene, and the active ingredient was the *Bt* protein which was pesticidal, hence it came under the EPA's jurisdiction.

In the US, we were a small team of three permanent employees, myself, a protein biochemist, Demi Vlachos, a toxicologist with experience in agrochemicals, and Rich Lotstein, a molecular biologist. I was responsible for the lab work and was on loan from one of the research groups where I had been responsible for a small group dedicated to identifying novel insect control agents; Demi was responsible for the submission and Rich was head of Regulatory and Government Affairs. I was allowed to hire two temporary scientists, Devon Brown and Trisha Fearing. Devon was a trained medical technologist with molecular biology experience and Trisha was a new graduate from NC State University. We were a tightly knit team feeling our way and making decisions on what we thought would make the world willing to allow their kids to eat corn chips made from our product. Our primary goal was approval in the US, with secondary goals of approval in the EU, Japan and Canada.

Suddenly, our work was the highest priority effort on-going at CIBA's biotech center. We were in a race against Monsanto and Sandoz and the stakes were pretty high. It took us two years to get the studies done, reports written, and submission made. It took multiple consultations with the agencies and a lot of guess work about what made sense, what questions needed to be addressed, how to address these questions, what was sufficient, how to prove a null, etc. We tried to organize our studies based on the practices used to evaluate the safety of an agrichemical. But we ended up producing six different text substances—for an agrichemical only one is used. We created new ecotoxicology study designs because the ones used for chemicals were inappropriate. For example, we used pollen as a test substance to be fed to ladybugs because the pollen contained the *Bt* protein and to us it made no sense to paint the *Bt* protein on the ladybugs thorax as is done with chemicals. Similarly, the Daphnia studies used pollen as the test substance. We used *Bt* protein purified from *Bacillus thuringeinsis* in toxicology and animal feeding studies not from the corn plant because the level of protein was so low we could not obtain sufficient quantities for a toxicity study. We also purified the PAT protein from an *Escherichia coli* overexpression system and conducted our acute oral mouse toxicity study with that even though we could not detect the PAT protein in plant tissues at all (due to the fragmented promoter). We had to come up with a way to assess the potential allergenicity of the newly introduced proteins. This included developing methods to use simulated mammalian gastric and intestinal fluids as described in the U.S. Pharmacopoeia which defined these fluids but did not specify volumes or protein concentrations to use, times, etc. A lot of development work was required as no study guidance was in place. However, we were really fortunate to be working with a product that gave a 15% higher yield when the pest was present and had no impact on yield when the pest was absent. There was no impact of genetic background and no impact on agronomic performance or composition. It was simply corn as we knew it that now could protect itself from ECB without the use of pesticide.

A requirement for any pesticidal product is to have critical information on the product label (in our case the seed bag label) including the maximum amount of active ingredient that is allowed or would be expected to be present at any one time. Hence, we needed to know how much *Bt* protein would be present at any time during the growing season and in which part of the plant—in case, for instance, if a hail storm came and the crop needed to be plowed under, what would be the exposure to soil organisms. This information also allowed us to calculate safety margins that were tested in our toxicology and ecotoxicology studies.

The submission was hand delivered to the EPA by Demi and Rich on Demi's birthday in July 1994. Then a year of waiting and answering questions and making submissions to the USDA, FDA and other agencies around the world before EPA approval was received in 1995 in time to sell the product for planting in 1996. The approval included an exemption from tolerance meaning that there was no maximum level of exposure to the *Bt* protein that would be considered unsafe. Deregulation from the USDA and approval from the FDA came also in 1995. Twelve years after the formation of the biotechnology unit by CIBA-GEIGY, two years after our ten year deadline, we had our first product ready for market and the focus groups determined a good name would be Maximizer™ Corn Hybrids containing Knockout control (built in corn borer control).

Global Issues

We had assumed that once we had approvals from Europe, Japan would follow suit. Little did we know that Europe would end up being a bigger road block than Japan. With regard to our application, it was submitted using France as the rapporteur country, in accordance with Directive 90/220/EEC. The rapporteur country was selected based on a couple of factors relating to the relative support of the product. France would have been a potential production country for this product and was advanced scientifically in the biotechnology area. Now, they are one of the strongest anti-GMO advocates. Nevertheless, back in the nineties our application was made via France. Patricia Ahl-Goy was our person on the ground in Europe to support and defend our dossier. Europe wasn't too keen on the idea of using the *Bt* protein purified from *B. thuringiensis* and not from the plant in our safety studies, so we prepared an enriched fraction from maize leaves (enriched $40\times$ but still <0.1% of the protein) for another acute oral mouse study. This proved acceptable to Europe but the EPA rejected this study. The timing in Europe was not ideal, although approval was received in late 1996 it was so controversial that a "moratorium" on biotech plants was imposed in Europe to allow them to prepare a major revision of their processes and policy. (This nicely delayed approval of our competitor's product which was great for us.) However, the new policies also included a very conservative and controversial labeling policy. Europe had experienced several food security issues the biggest ones were mad cow disease (1996), dioxin contamination in animal feed (1999) and a carbon dioxide issue in Coke products (1999) during this time that led to an overhaul of their system, new regulations and the formation of the European Food Safety Agency (EFSA) in 2002. EFSA has issued more than 10 guidance documents in the ensuing ten years and these together with the so-called Implementing Regulation that went into effect in December of 2013 have greatly raised the hurdles and constricted the pathway towards approvals of biotech products in Europe (Devos et al. 2014). Only a couple of European countries actually grow any biotech products and only a couple of products are actually approved for cultivation there (James 2014). Trade disputes and changes in practices have resulted. However, since Europe is dependent upon the import of soybean to support its livestock industry, import approvals have been granted.

As we were awaiting approval from Europe, we were responding to questions from Japan as well. It was amazing the level of detail required from a country where the product wasn't going to be marketed or planted. The review was based on the possibility that when maize grain arrived in Japan some might fall off of a truck and germinate on Japanese soil. This all seemed pretty far-fetched, nevertheless, all studies were submitted and a thorough evaluation was made. In the end, we received approval in 1996 prior to that from Europe!

Canada was also a key country on our radar. The Canadians took a very pragmatic approach and have designed a system that evaluates novel plants regardless of the process in which they were produced. In other words, they did not care if the plant was a result of genetic engineering, conventional breeding or mutation breeding, all are evaluated in a similar manner if the crop had never before been encountered in Canada. I actually had the honor of preparing the "Plant with Novel Trait" application. Approval was received for use as food in 1995, for feed and cultivation in 1996.

It is Never Over

Once approvals were granted, new issues arose all of which had to be addressed and some came under the Sect. 6(a)(2) Adverse effect clause of FIFRA which requires pesticide product registrants to submit adverse effects information about their products to the EPA. Basically, if information relating to the safety of the product arises either publically or privately, this information and a discussion regarding the impact on the use of the product must be reported. Examples that we ended up having to evaluate included a grower who noted that when he released his cows to graze in his corn fields—more visited the conventional field than the Bt field. It turned out that there was more grain on the ground (due to plant lodging from insect damage) in the conventional field than the Bt field, so the cows were going to where the food was. Another grower complained that the stubble remaining in a Bt field caused more tire issues for his farm equipment than the conventional field. There was much less fungal disease present in the Bt corn stubble than the conventional stubble because there was less insect damage for the fungal pest to use to invade. Actually, an unintended benefit of Bt corn is the reduction of mycotoxin contamination because there are less fungal infections due to reduced insect infestation (Wu 2006). A study from France indicated reduction of Dipteran insects in Bt corn fields than in conventional corn. The question was is the Bt protein not as specific as originally thought? However, the explanation for the observation was found to be a result of reduced fras (excrement) from caterpillars (such as ECB) in the field and hence, no place for the Dipteran insects to lay their eggs.

The most serious issue was the Monarch butterfly report in 1999 from John Losey, an assistant professor at Cornell University. He claimed that the pollen from Bt corn when landing on the milkweed plant would provide a pathway by which the Monarch butterfly, a lepidopteran insect would be exposed to the Bt protein. Needless to say, this was a major problem for our Bt corn product. The field of ecological entomology was revitalized as research money was thrown to everyone from everywhere to figure out the true impact. Monarchs are not pests of corn; they are 100% dependent on milkweed as their food source. Milkweed plants may grow in or around corn fields, but in order to be exposed would need to be coated with Bt pollen only from our event (other events did not express Bt in the corn pollen), which only is produced during a 2–3 week period during the growing season. Monarch larvae are sensitive to the Bt protein. At the end of the day, it was concluded that more Monarch butterflies are killed by windshields than by Bt corn pollen. Nevertheless, other *Bt* corn events that did not express *Bt* in their pollen were found more acceptable and in reality were more efficacious. So this original *Bt* event (Bt176) is no longer marketed and was replaced by other products.

Status of Regulation Today

In this day and age, we are a global community which means that not only must our products be considered by US regulators but also by the entire globe. The World Health Organization (WHO) and the Food and Agriculture Organization of the United Nations (FAO) established in 1963 the Codex Alimentarius to set International Food Standards. Codex, as it is referred to, "is about safe, good food for everyone—everywhere." Currently there are 185 Codex Members and 1 member organization (EU). Most countries with official regulations on genetically modified organisms base their regulations on the Codex requirements (Codex Alimentarius 2003; Privalle et al. 2012; Bartholomaeus et al. 2015).

Today, safety assessments of biotech products take a multipronged approach considering both the safety of the gene, the gene product and source of the gene; as well as the safety of the crop containing the new gene (and expressing the protein or exhibiting the trait of interest)—how comparable is this crop to the crop without the new gene (the conventional crop). To evaluate the safety of the gene, the source of the gene is taken into account. Theoretically, any organism can be a donor organism or source of the gene. The whole advantage of biotech crops is that the source of the gene is not restricted to an organism that is sexually compatible with the crop of interest. However, due to public acceptance issues, there was a decision taken to not source any genes from animal sources. As a result, most genes are taken from either plants, algae, fungi, or microorganisms. Examples of organisms besides animals that would not be considered good sources would include an allergenic food such as peanut or disease causing microorganism such as *Bacillus anthrax*. Evaluation of the safety of the gene product is relatively easy to predict using bioinformatics approaches. The databases of protein sequence information that are publically available are easily screened using algorithms developed precisely to predict how closely related proteins are and aid in the prediction of function. These databases grow daily so avoiding known bad actors is easy to accomplish. Furthermore, there are specific allergen databases such as allergenonline.com that allow more focused searches (Ladics et al. 2011; Goodman et al. 2008). Criteria laid out by regulatory agencies and by the Codex Alimentarius are highly conservative such that even though only a minute fraction of the proteins are actually allergenic about 15% of all proteins would be tagged as potential allergens when doing these kinds of searches.

In addition to the bioinformatics assessment, the safety of the protein is directly assessed using purified protein. This protein is usually produced in a recombinant bacterial system as it is extremely difficult to produce sufficient protein from the crop plant itself. One of the first studies that is conducted is the demonstration that the protein produced in the microbial system is similar to that produced by the crop. Characterization of both physical and function parameters are involved. A complete description of the safety assessment of the protein is described by Delaney et al. (2008) and Hammond (2008). Examples for specific proteins are described in Raybould et al. (2013), Privalle et al. (2000), Hérouet et al. (2005) and Madduri et al. (2012). Future products may involve proteins that are more intractable, for these types of proteins slightly different approaches may be necessary (Bushey et al. 2014). According to Delaney et al., the safety assessment of the protein is divided into two tiers. The first tier includes an assessment of the protein's history of safe use, bioinformatics analysis, digestive fate or stability assessment, and expression level determination (to assist in determining exposures). The second tier involves acute and subchronic toxicity tests usually using mice as the test species. In a standard risk assessment approach, such as used for agrichemicals, the second tier studies are only if there are indications from the first tier studies that additional studies are warranted. In the biotech world, several second tier studies are conducted regardless of the outcome of the first tier and are required in some geographies.

The other half of the safety assessment of the biotech product is focused on the safety of the crop. Several questions are addressed—does the crop perform in a manner similar to its conventional counterpart—except for the newly imparted trait? What is the molecular organization of the insertion site—including where did it insert, how many copies were inserted, were any genes interrupted, did any vector backbone get inserted, what is the sequence of the insert? Is the nutrient or antinutrient composition of the crop altered? Is the trait inherited in a Mendelian fashion? Is the performance stable across generations? Where there any unintended effects? What is the environmental impact of the crop? Has the wholesomeness of the crop been impacted?

Benefits and the Future

A result of all these levels of assessment is that food produced using crops enhanced through biotechnology are the most studied and understood foods consumed. However, they are extremely controversial and are politically charged. Most people simply do not understand or appreciate where their food comes from. On the other hand, crops produced with these technologies demonstrated an extremely rapid and high acceptance rate by growers (Fig. 3). Currently 28 countries cultivate these crops (James 2014) and both small and large growers enjoy the multiple advantages they provide. Benefits to the world are largely environmental—topsoil is preserved, production from current arable acreage is maximized, pesticide applications are greatly reduced, less petroleum is used as growers do not need to enter their fields as often; beneficial insects are not impacted; biodiversity is unimpacted; mycotoxin contamination is reduced and yields are increased. Taken all together, this is a very "green" technology.

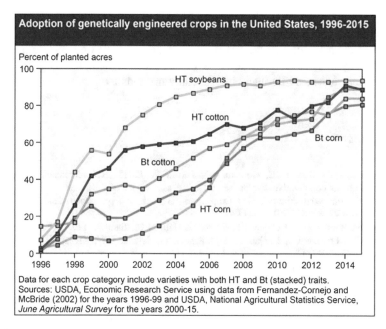

Fig. 3 Adoption of biotech products in the US (http://www.ers.usda.gov/data-products/adoption-of-genetically-engineered-crops-in-the-us/recent-trends-in-ge-adoption.aspx)

So with all these great benefits why are regulations increasing instead of decreasing? Why are consumers so wary of these products? Is this a simple matter of education? The first generation of biotech products was designed to solve problems for the grower—the so called input traits. We spent a lot of time and effort interacting with the growers to educate and be educated about agricultural challenges and possible technology-based solutions. We also spent a lot of time talking to the down- stream users—the dairy cattle association, the poultry associations, the cattlemen's association, the pet food producers. I got to go all over the US and make presentations at the meetings of these groups. It was fascinating. Strategically, the error was made in not doing the same with the consumers.

The second generation of products was going to be crops that had traits of value to the consumers—but where are they? Technically, making crops with improved nutritional attributes and other "output traits" was possible (and has been done) but the cost for obtaining global approvals is huge and often with the smaller crops cannot be justified. Currently, in the US only nine crops have biotech products actually on the market.

As we look forward to the population growth that is predicted for the world and we consider how food, fiber, energy must be available to support this growth and also consider water limitations, land limitations and climate change, there is no question that sustainable agriculture is absolutely required. Ecology tells us that populations can only be sustained as long as their food supply. Without the smart use of our resources, our food supply will be finite. Human beings are endlessly inventive yet somewhat precautionary. I see that the concept of sustainable agriculture is the result of a realistic evaluation of the resources we have available and the best way to use them. One key facet of sustainable agriculture has got to be the use of crops enhanced through biotechnology methods.

References

Bartholomaeus A, Batista JC, Burachik M, Parrott W (2015) Recommendations from the workshop on comparative approaches to safety assessment of GM plant materials: a road toward harmonized criteria?, GM Crops Food Biotechnol Agric Food Chain 6:2, 69–79. doi:10.1080/21645698.2015.1011886

Bushey DF, Bannon GA, Delaney BF, Graser G, Hefford M, Jiang X, Lees TC, Madduri KM, Pariza M, Privalle LS, Ranjan R, Saab-Rincon G, Schafer BW, Thelen JJ, Zhang JXQ, Harper MS (2014) Characteristics and safety assessment of intractable proteins in genetically modified crops. Regul Toxicol Pharmacol 69:154–170

Codex Alimentarius. Guideline for the conduct of food safety assessment of foods derived from recombinant-DNA plants (CAC/GL 45-2003). Codex Alimentarius Commission. Rome, Italy: Joint FAO/WHO Food Standards Programme; 2003

Delaney B, Astwood JD, Cunny H, EichenConn R, Herouet-Guicheney C, MacIntosh S, Meyer LS, Privalle L, Goa Y, Mattsson J, Levine M (2008) Evaluation of protein safety in the context of agricultural biotechnology. Food Chem Toxicol 46:S71–S97

Devos Y, Aquilera J, Diveki Z, Gomes A, Liu Y, Paoletti C, du Jardin P, Herman L, Perry JN, Waigmann E (2014) EFSA's scientific activities and achievements on the risk assessment of genetically modified organisms (GMOs) during its first decade of existence: looking back and ahead. Transgenic Res 23:1–25

Federal Insecticide, Fungicide, and Rodenticide Act (FIFRA) (1972) 7 U.S.C. §§ 136–136y

Food, Drug, and Cosmetic Act (FDCA) (1938) Pub. L. No. 75–717, 52 Stat. 1040

Goodman RE, Vieths S, Sampson HA, Hill D, Ebisawa M, Taylor SL, Van Ree R (2008) Allergenicity assessment of genetically modified crops—what makes sense? Nat Biotechnol 26:73–81

Hammond BG (2008) Food safety of proteins in agricultural biotechnology. CRC Press, Boca Raton

Hérouet C, Esdaile DJ, Mallyon BA, Debruyne E, Schulz A, Currier T, Hendrickx K, van der Klis RJ, Rouan D (2005) Safety evaluation of the phosphinothricin acetyltransferase proteins encoded by the *pat* and *bar* sequences that confer tolerance to glufosinate-ammonium herbicide in transgenic plants. Regul Toxicol Pharmacol 41:134–149

James C (2014) Global Status of Commercialized Biotech/GM Crops: 2014 ISAAA Brief No. 49. ISAAA, Ithaca

Koch MS, Ward JM, Levine SL, Baum JA, Vicini JL, Hammond BG (2015) The food and environmental safety of Bt crops. Frontiers Plant Sci 6:283. doi:10.3389//fpls.2015.00283

Ladics GS, Cressman RF, Herouet-Guicheney C, Herman RA, Privalle L, Song P, Ward JM, McClain S (2011) Bioinformatics and the allergy assessment of agricultural biotechnology products: industry practices and recommendations. Regul Toxicol Pharmacol 60:46–53

Madduri KM, Schafer BW, Hasler JM, Lin G, Foster ML, Embrey SK, Sastry-Dent L, Song P, Larrinua IM, Gachotte DJ, Herman RA (2012) Preliminary safety assessment of a membrane-bound delta 9 desaturase coandidate protein for transgenic oilseed crops. Food Chem Toxicol 50:3776–3784

National Environmental Policy Act of 1969 (NEPA), Pub. L. 91–190, 42 U.S.C. 4321–4347, January 1, 1970, as amended by Pub. L. 94–52, July 3, 1975, Pub. L. 94–83, August 9, 1975, and Pub. L. 97-258, § 4(b), Sept. 13, 1982)

Plant Variety Protection Act (PVP): Public Law 91–577, 84 Stat. 1542–1559; Dec. 24, (1970) as amended by Pub. L. Law 96–574, 94 Stat. 3350–3352. Sec. 1–19; Dec. 22, 1980; Pub. L. 97–164, 96 Stat. 37–38, 41, and 45. Sec. 127, 134, 145; April 2, 1982; Pub. L. 100–203, 101 Stat. 1330–28. Sec. 1505; Dec. 22, 1987; Pub. L.102–560, 106 Stat. 4231, 4232. Sec. 3; Oct. 28, 1992; Pub. L. 103–349, 108 Stat. 3136-3145. Sec. 1–15; Oct. 6, 1994; Pub. L. 104–127, 110 Stat. 1186, Sec. 913; April 4, 1996 and Regulations and Rules of Practice, 7 CFR, Part 97, as of September 1, 1996; Federal Register: August 2, 2000 (Volume 65, Number 149, 47243–47245); Federal Register: January 10, 2003 (Volume 68, Number 71, 1359–1360); Federal Register: May 19, 2005 (Volume 70, Number 96, 28783–28786); Federal Register: September 16, 2005 (Volume 70, Number 179, 54609–54612)

Privalle LS, Wright M, Reed J, Hansen G, Dawson J, Dunder EM, Chang Y-F, Powell ML, Meghji M (2000) Phosphomannose isomerase—a novel system for plant selection. Mode of action and safety assessment. In: Fairbairn C, Scoles G, McHughen A (eds) Proceedings of the 6th international symposium on the biosafety of genetically modified organisms, University of Saskatchewan Press, Saskatoon, Canada, pp 171–178

Privalle LS, Chen J, Claper G, Hunst P, Spiegelhalter F, Zhong CX (2012) Development of an agricultural biotechnology crop product: Testing from Discovery to commercialization. J Agric Food Chem 60:10179–10187

Raybould A, Kilby P, Graser G (2013) Characterising microbial protein test substances and establishing their equivalence with plant-produced proteins for use in risk assessments of transgenic crops. Transgenic Res 22:445–460

Sanford JC, Klein TM, Wolf ED, Allen N (1987) Delivery of substances ionto cells and tissues using a particle bombardment process. Part Sci Tehchnol 5:27–37

Wu F (2006) Mycotoxin reduction in Bt corn: potential economic, health, and regulatory impacts. Transgenic Res 15:277–289

Trails and Trials in Biotechnology Policy

Jennifer Kuzma

A Natural Scientist

It was the mid-1980s when my career in biotechnology began at a smallish, Catholic liberal arts college in the Upper Midwest. I knew that I had to major in natural science, as those were the terms of my full-tuition scholarship for 4 years. I had an affinity for chemistry and math; however, during my freshman biology class, I learned about how the natural world operated at the molecular level, and my interest was piqued. Soon thereafter, I joined the laboratory of a young, cell and molecular biology professor who had just joined the college and was brave enough to start a research program with only undergraduates. Our college did not have many other research labs, as we were an institution with just budding strengths in the natural sciences. However, in this environment, I was able to get excellent training in many facets of research, unguided by a cadre of graduate students and postdocs. For example, there were four students in my molecular biology class, and I remember reading the peer-reviewed technical literature about PCR (which had just arrived on the scene) and debating the molecular methods with my classmates. Overall, it was a stellar education.

Our lab, under the direction of Professor Jennifer Cruise (my first, fabulous female mentor) at the College (now University) of St. Thomas in St. Paul, MN, I worked on molecular, cellular, and biochemical mechanisms of liver regeneration, and the lab subjects were usually rats. My adviser knew and respected that I would not be killing any rats (not that I am or was against it, but I just simply couldn't stomach it); however I was asked to grow up viruses for our work to deliver molecules into cells in order to study biochemical pathways. I still remember the day when I tried to harvest the virus from eggs and the needle got stuck (or was grabbed?) by a chicken fetus. I ran out of the lab, practically screaming. I continued

J. Kuzma (✉)
School of Public and International Affairs, Genetic Engineering and Society Center, North Carolina State University, Raleigh, USA
e-mail: jkuzma@ncsu.edu

© Springer International Publishing AG 2017
L.S. Privalle (ed.), *Women in Sustainable Agriculture and Food Biotechnology*,
Women in Engineering and Science, DOI 10.1007/978-3-319-52201-2_6

to work in the lab, but I didn't have to harvest viruses from eggs again, and at that point, I was pretty sure that I should look for graduate work that would focus on plants or microbes.

During this time, between courses and lab work, I was politically active. In the 1980s, key issues were apartheid in South Africa and the proliferation of nuclear weapons for deterrence in the Cold War with the Soviet Union. I was also a part of a social justice group at the college, as movements to help the impoverished and oppressed and work for peace were prominent in liberal Catholicism at the time. I had many interests in areas outside of the natural sciences, and I was able to take several courses to nurture these interests. I had a passion for philosophy and ethics, music, and social science. However, because natural science and math came easy to me and I was also interested in them, my advisers encouraged me to continue this path in graduate school given the need for more women in these male-dominated areas.

The decision to go to graduate school was somewhat capricious. I came from a family with 6 kids, and my father was the first to go to college in his 12-sibling family. He was a veteran of World War II and went to the University of Minnesota on the GI bill, graduating in chemical engineering. Although he didn't have a generous salary working as an engineer in the iron ore mines in Northern Minnesota, he and my mom made sure all of their 6 kids went to college. But they also made sure that my siblings chose sensible, practical careers in fields like dentistry or health care. We were not dirt poor, but we didn't have much money either, so when my college adviser suggested that I should go to graduate school and get my Ph.D., my first question was whether I would have to pay for it. I didn't really know what graduate school or a Ph.D. was about at the time. He said I'd make enough money as a graduate research assistant to live on and go out to a movie once in a while. That was enough for me. So as a junior in college, I started thinking about graduate work in biochemistry to blend my interests in both organic chemistry and biology. I majored in both chemistry and biology, as my college didn't have a major in biochemistry at the time. Biochemistry was considered interdisciplinary.

Around the same time, I started reading and thinking more about the philosophical, ethical, and social questions surrounding genetic engineering. It was during the time of the first proposed field-trials of genetically-engineered organisms (GEOs) in the mid-1980s, the ice minus bacterium. I found a lab at the University of Colorado Boulder working on ice minus bacteria. The lab also worked on the isoprene bio-synthetic pathway in plants and called itself an "environmental biochemistry" group. The focus on plants and bacteria ensured that no rats or chicken fetuses would be in my future. UC Boulder also had a strong program in biochemistry with several prominent faculty pioneering RNA catalysis, DNA synthesis, and other work. Finally, Boulder was a good cultural fit for my social justice leanings.

Graduate school proved to be a trying time with most of my experiments destined to fail. The lab was a pleasant place to be people-wise, but our work on big questions at the nexus of the environment and biochemical pathways and exploration of previously undiscovered proteins meant that there were no easy experiments. I did manage to make progress, however, and one of my contributions was the discovery that bacteria produce the volatile compound isoprene. Isoprene is a precursor to natural rubber and is usually obtained by cracking petroleum. With bacteria producing quite a bit of relatively pure isoprene gas, we envisioned a more sustainable future source. I obtained a patent on that discovery with my thesis adviser Prof. Ray Fall and the lab technician Michele Nemecek-Marshall. Unfortunately, we were about 15 years ahead of our time with our interest in more sustainable bioproducts, and the patent sat for many years. It was not licensed until over a decade later, in 2007. However, it was satisfying to know I was working on something with broader societal and environmental implications, and today, Goodyear and DuPont Industrial Biosciences are making BioIsoprene™ for the production of rubber to help reduce the tire industry's dependence on oil-based products. During my thesis work, I also purified an enzyme for isoprene synthesis and tried to clone the gene, but that was proving to be difficult given that genomic science was not really a part of the early 1990s. Thankfully, I eventually had enough work for a thesis, so I moved on.

My second contribution in natural science was the discovery of abscisic acid in a signaling pathway for plant responses to stress (drought, salinity) as a postdoc at Rockefeller University in New York City. The move from Boulder to the Upper East Side of NYC was quite a cultural shift, and this was a hard time for me in more ways than one. Lab work was becoming more tedious and less interesting to me, and I would escape the lab and "concrete jungle" of Manhattan many weekends to play in beach volleyball tournaments on the Jersey Shore or Long Island, where I eventually met my spouse at a charity tournament. Despite the growing disdain for lab work, I did feel proud of the bigger picture of my work, as the ultimate goal would eventually be to engineer plants tolerant to stress, so that the hungry and suffering would have more food. The results of my work were my first (and only technical) publication in the journal *Science*. I wrote the first draft and most the article, but because the professional technician did most of the hours on the wet-lab work, I relinquished first authorship to her. At that point, my passion for a career in biochemistry and molecular biology was waning and it didn't really matter to me. So with a patent and high profile publication in hand, I finally allowed myself to acknowledge just how unhappy I was with the day-to-day work of laboratory science. It was just not for me. The focus on one enzyme, biochemical pathway, or gene was too detailed for me, yet I retained a deep and strong passion for the broader societal context and implications of biotechnology, so looked for ways to move on using my Ph.D. and postdoctoral experience. At this time, I had no idea how I could blend my interests in biotechnology with social science, politics, and ethics.

A Risk Analyst

Thankfully, I soon thereafter found an advertisement for a science policy fellows program and applied on a whim. On crutches, suffering from a volleyball tournament injury, and with my ACL ligament freshly repaired, I hobbled to Washington DC in early 1997 for an interview for the American Association of the

Advancement of Science (AAAS) Science Policy Fellows program. I remember that I had written my policy brief for the interview on a subject I knew little about, the mining of methane from the ocean floor as a more sustainable fuel source. It was of interest to me because this technology potentially posed both significant environmental risks and benefits. I remember that the interview panel asked why I would write on something that I didn't know much about, as opposed to something I knew more about like biochemistry or molecular biology. My answer must have been OK, or they took pity on me because of the crutches and full-leg brace, as I landed the AAAS fellowship. It is true that in the world of policy-practice in Washington DC, you have to get up to speed on diverse issues in a fairly short time. This was a perfect fit for someone like me with broad interests at the nexus of science and society, but also deep analytical skills.

For the AAAS fellowship, I was placed into a topical area that I knew little about too, risk analysis for food-borne pathogens and hazards. So the first few months in Washington DC at the US Department of Agriculture, I mainly (and smartly) kept my mouth shut and soaked in the technical information and nuances of the politics. I observed where natural science and decision making intersect—where the rubber hits the road so to speak–in risk analysis, decision making, and regulatory policy. I owe a great deal to another fabulous female mentor, Dr. Nell Ahl, for giving me the chance to learn risk analysis methods, regulatory policy, public policy, and politics of decision making during my fellowship. She was patient with me, as a novice to these worlds, and in time, I do think I contributed to the office she directed. But mainly, I learned.

One of the highlights at the USDA was writing and helping to negotiate the interagency politics over the scope of a risk assessment for mad cow disease in the United States, a big concern in the late 1990s. Different agencies within USDA had distinct priorities and missions. One unit wanted the risk assessment to stop at the quantification of the risk of animal disease, and yet another wanted it to go all the way to estimating human health outcomes. Diplomacy was needed to see both sides and reflect compromises in writing. Another big project for me was helping with a farm-to-table risk assessment for *E. coli* 0157:H7 in ground beef (Ebel et al. 2004). I worked on the slaughter module, helping to model contamination in animal processing plants. The same person who couldn't kill rats was now being asked to visit slaughterhouses. It wasn't pleasant, but it was interesting, and in addition to learning about potential sources of contamination with the bacterium, I developed an appreciation for the hard labor that workers in these plants do.

Foodborne contamination with *E. coli* 0157:H7 was a big issue at the time. I remember my first regulatory policy meeting on this topic in DC. For most of the day, natural scientists risk assessors, and regulatory policy experts talked about how very low the risk of death or severe illness was from eating ground beef. Stakeholders from cattle or beef industry associations argued against a stringent standard for the pathogen in ground beef. Then, in the closing panel, a parent whose child died because of the Jack in the Box *E. coli* 0157:H7 outbreak in hamburger stood up and made it clear that even one death was too much. This event wasn't too many years after my 1 year old nephew died in a drowning accident in the presence of a baby sitter, so needless to say, her plea really touched me, as well as others in that room. So much for "sound science" in telling us what is probable or right.

The fellowship year was transformative, and many lessons were learned from my observations and work. One was that natural scientists, especially those with a stake in the issue or technology, often display hubris by claiming to have the answer and know what is safe or right for all people, when in fact there is a great deal of uncertainty and interpretation of evidence that comes into play. For example, the agricultural minister in the UK fed his child a burger on TV during the height of the mad cow disease crisis and made claims that there was no risk to humans, or no link between the animal form of the disease (BSE) and the human illness (nvCJD) (Leiss and Powell 1997). He turned out to be wrong, causing considerable loss of trust in the UK for regulatory policy officials. It didn't help the situation that nvCJD is always fatal and a horrifying neurological disease.

I learned how assumptions and values color even the best of the risk analyses used for decision making. Not only was there intentional bias in the political sphere, but also unintentional bias or world views that cause even the best and brightest natural scientists to make strong claims about uncertain situations. Mainly, I came to understand how I, as a natural scientist and technologist, did not have the answers to saving the world from hunger or petroleum dependency. For example, sometimes producing a bio-product takes *more* oil than it replaces, and sometimes a GE crop will pose risks, however small, that are not acceptable to people in light of the fact that they do not receive the benefits. I learned that technologies, which come with their own risks, are not always (or even usually) the best way to address global problems, which are caused by a confluence of natural, social, economic, and political factors. More often, social and political systems are the main causes of hunger and petroleum dependency.

A Science and Technology Policy Practitioner

Approaching 30-years old now and married, it was time to find a "permanent" job. So after the fellowship in late 1998, I applied for and received a position at the National Academy of Sciences (NAS) (now National Academies of Science, Engineering, and Medicine NASEM) as a study director in the area of biotechnology for the Board of Biology (now Board on Life Sciences). I didn't really have a boss when I first arrived, as the board was in transition. So, I got to know the President of the NAS, Professor Bruce Alberts, quite well in the first year, as he was most interested in NAS studies related to biotechnology as a pioneer in molecular biology. I was excited to be back in the area of biotechnology after my foray into microbial food safety. It was a time during which the Coordinated Framework for the Regulation of Biotechnology had been formalized through proposed or enacted regulatory rules (OSTP 1986; USDA 1993, 1997); GE crops were proliferating in field trials, and they had recently entered the open market (Kuzma 2013).

My first big assignment at NAS was as study director for a committee report examining the science and regulation of GE plants designed for pest-protection. Instead of spraying pesticide on a field, these crops had pesticide-like proteins within them, mostly from the bacterium *Bacillus thuringiensis*. The study was called for by the NAS membership itself, as the Environmental Protection Agency (EPA) had proposed to regulate GE plants with Bt proteins and other plant pesticides (EPA 1994, 2001), and several NAS members (prominent molecular biologists, biochemists, plant pathologists, or agronomists) were not too happy about this situation.

The USDA already regulated GE plants under the Federal Plant Pest Act (FPPA) (USDA 1993, 1997). USDA's mission as an agency is to protect U.S. agriculture, and through the FPPA, it looks at risks to agriculture from plant pests. In the case of GE crops, the plant pest sequences used for genetic engineering, like *Agrobacterium* and Cauliflower Mosaic Virus, were used as the regulatory hook for USDA. In contrast, the EPA has the mission to more broadly protect the environment. EPA proposed to exercise authority under the Federal Insectcide, Rodenticide, and Fungicide Act (FIFRA) for GE plants with pesticidal proteins or molecules (EPA 1994, 2001). Under FIFRA, EPA takes a rigorous look at safety to non-target organisms and requires significant data prior to approval for marketing a pesticide. It also has post market re-registration and monitoring authority. Under the FPPA (now the Plant Protection Act, PPA), USDA's assessment focuses on "plant pest" risks to agriculture, and once a crop is deregulated, no post-market monitoring authority exists. EPA's proposed role under the CFRB would include assessments for impacts on non-target insects, birds, fish, and mammals from GE crops, as well as the human safety of ingesting residues of plant pesticides like Bt in food.

The NAS molecular biologists did not want any additional regulation of GE crops, and therefore asked the NAS to commission a study in the hopes that the committee would come out against EPA regulation (and perhaps even question USDA's regulation). Many of them felt that GE crops should not be singled out for regulation at all, as conventionally bred crops can pose similar risks. A 1987 NAS committee on which several of those same NAS members served "Introduction to rDNA Organisms in the Environment" stated that the risks of GE crops and conventional crops are "the same in kind" and that there are "no new categories of risk" (NAS 1987). The "science-based" conclusion to them was that therefore, there should be no formal regulation of GE crops, just like conventional crops.

It was a contentious study and we were criticized by both the pro-GE and anti-GE groups—we had a few committee members from private consulting groups for biotechnology industries, so we were criticized by NGOs for being biased towards industry from the start of the study. The pro-GE groups criticized us for including scientists with ties to environmental NGO groups and for not including some of those NAS members that called for the study on the committee. On the morning of the report's release, protestors surrounded the building dressed in lab coats or as Monarch butterflies, as a recent study had just showed that Bt pollen was toxic to Monarch larvae in laboratory feeding studies (Losey et al. 1999). Former Presidential candidate, then Representative Dennis Kucinich, was outside the building with the protestors, some who were holding signs with slogans like "Just Say No to GMOs". It was an intriguing experience for my first report at the NRC as study director.

In the report, the committee extensively discussed a strict "scientific basis" for a regulatory system and decided it was not feasible. For such a foundation, two logical policy options existed—if GE crops were equivalent to conventional crops, we should regulate both conventional and GE crops or regulate neither. These two extreme options were not respectively practical or protective of ecological or human health. Furthermore, choosing between one or the other option is not a decision solely based on "sound science", as one's values must guide whether everything (GE and conventional) should be assessed and regulated prior to or whether nothing should be regulated. The decision between those two "science based" options involves world views about the role of government in protecting ecosystems compared to the role of technological and economic development. The committee also disappointed the NAS members calling for the study by supporting a role for the EPA in regulating GE crops, given that USDA's scope was limited to agricultural protection and that EPA focused more broadly on ecosystem harm. The committee also suggested that given the lack of experience, uncertainties, and public concern associated with GE plants, it made sense to regulate them and not conventionally bred crops for the time being (NRC 2000).

Through these discussions, I learned that it is impossible to design a completely "science-based" regulatory system. Regulatory capture can be informed by natural science, but judgements come into play depending on levels of uncertainty, novelty, potential harms or risks, and other societal concerns (Kuzma 2016a). Regulation should also be informed by social science, values people hold, and ethical criteria. Science can help to tell you what is, but cannot dictate what to do.

I observed the different communities associated with genetic engineering during my time at the NAS. Although I do not want to rearticulate and support all my arguments on the subject of GEOs governance (see Kuzma et al. 2009; Kuzma 2013, 2016a as overviews), I have learned that natural scientists (more specifically, those natural scientists on the technology development side) call for "evidence-based" and "science-based" regulation, but still come with just as many value-based arguments and biases as those who prefer more precaution (consumer and environmental groups, many toxicologists and ecologists, several social scientists) before releasing GE crops in the field.

One indication of the biases was in the world of peer-reviewed publication on risk science associated with GE foods. In this domain, I also observed how those who published studies that showed any potential harm from GE crops in the peer-reviewed literature were discredited, pressured to retract the papers, and their findings dismissed by the mainstream plant biotechnology community (e.g. the Puzstai, Chapela, and Seralini cases as discussed in Bardocz et al. 2012; Loening 2015). When studies do not find risks with GEOs, protocols and designs of them are not of concern to industry and academic product developers or GEO advocates; however, studies that use similar or the same designs that show potential harms are harshly critiqued and met with vile (Meyer and Hilbeck 2013; Hilbeck et al. 2015).

Concerning the critics of studies that suggest potential risks or harms, Loening (2015) states that "The vehemence of the(se) critics is not matched by their evidence; it is often based on entrenched assumptions and on mis-representing published material. The arguments have challenged normal healthy scientific dialogue, and appear to be driven by other motives." I agree, even though I personally do not think GE foods currently on the market pose significant danger to human health (and many a GE food can be found in my kitchen and is eaten by my kids in a given day). However, I am dismayed by the unscientific approach taken by critics on both sides of the safety debate. As someone with risk assessment, biochemistry, and policy backgrounds, I believe that we, as "technological elites" are not communicating honestly. No one should be claiming categorically that "GE foods are safe"; just as no one should be claiming that "GE foods are dangerous". Safety involves verifying an absence of an effect, which under logic rules does not constitute "proof". In other words, just because you see "no effect", it doesn't mean that there *is* no effect in a given study. *All* studies come with uncertainty, and the short term 90-d toxicity studies done on rodents for regulatory review, which are of insufficient time scales and unrealistic contexts, are no better categorically than the imperfect lifetime, whole-food feeding studies that may be imperfect. Some studies in the peer-reviewed literature do report biochemical and morphological changes that may indicate negative effects on test animals from consumption of GE foods (e.g. Dona and Arvanitoyannis 2009; Vecchio et al. 2009; Domingo and Bordonaba 2011; Carman et al. 2013; Bøhn et al. 2014; Gu et al. 2014; Glöckner and Séralini 2016; Lurquin 2016); but the balance of studies show no such effects (e.g. reviewed recently in Domingo 2016). The fact remains that we do not currently have a sound, scientific way to look for the long-term effects of a life-time consumption of GE foods in humans, especially if those effects are of the more subtle kind like food allergenicity, intolerances, or sensitivities. Let's be honest about this, and from a policy perspective, think more about the tradeoffs in potential risks and benefits of GE foods compared to alternatives. For tradeoff analysis, values must come into play and therefore in a democratic society, these conversations require societal dialogue and input. In fact, values come into play in risk assessment too (Kuzma and Besley 2008). Even with good data, although you can estimate a dose response curve, there is always uncertainty about where you draw the line for a regulatory standard, and therefore, of what is "safe" always comes with a value judgement (Kuzma 2016a, b).

I am a scientific and logical thinker, and am continually dismayed by the mainstream molecular biology community thinking that those who have objections to GE crops "just don't understand the science", "need to be educated", and don't get it or are "luddites". There is a lot of misinformation about GE out there, but not all those who are critical or questioning of GE crop-safety are doing so out of an agenda or ignorance. Also, most regular people I converse with can indeed understand the science with a little background and ask insightful questions. Let's give the critics some credit and have an honest exchange about tradeoffs and multiple types of criteria for making a decision about whether to use GEOs in a given situation or to address a particular societal problem.

Recently, at a national meeting, I questioned whether GE crops were needed to feed the world. In my opinion, this is a valid question given that we produce more calories per capita than needed (FAO 2016) and that GE crops have not consistently increased yields (NASEM 2016). I don't disagree that they may have a role under some conditions, but claims of needing them to feed the world are not scientific. They might be desirable, but they are not necessary.

For questioning the necessity of GE foods for food security, I was accused of wanting to let people starve by a developer of GE products at this meeting. Given my social justice background and political leanings, this comment infuriated me. No moral and sane person wants anyone else to starve. However, if we do have enough calories to feed the world, why doesn't it make more sense to focus on the socio-political distribution of food? Why are we not putting billions of dollars into those issues instead of GE crop research? This is a societal judgement based on values. It was questions like this in the practical science and technology policy world that led me to what will likely be the final phase of my career—a professor in policy and social sciences focusing on emerging technologies, governance, and decision making. In 2003, it was time to move back to academe, so I could study and write freely about these issues from the interdisciplinary perspectives that I accumulated.

A Professor in Policy and Social Sciences

My time as a professor for the past 13 years at two different universities, the University of Minnesota and North Carolina State University, has been spent on looking at questions of governance for emerging technologies, particularly nanotechnology, biotechnology, and synthetic biology. I am now a distinguished professor and endowed chair in a field in which I did not get my Ph.D. Sometimes that is a bit unsettling to me, as I do not have full legitimacy around hard core disciplinary social scientists. But at other times, it is a great asset to have a good foundation in both the natural and social sciences. It has well positioned me to try to cross barriers of understanding, different theories, and diverse methodologies. In the past decade, with colleagues and students, I have developed methodologies to help integrate diverse types of metrics (criteria) that can assist with decision making for emerging biotechnological products, to deal with uncertainty in risk governance, and to anticipate new products and their risk and benefit potentials well in advance of product development and regulation. I have also studied the intersection of values with evidence for genetic engineering and argued for middle-ground approaches to risk governance that blend precautionary and promotional perspectives (Kuzma 2016a, b) These are currently my main areas of contribution to the agricultural biotechnology debates.

It was a great honor in 2016 to be asked to serve on a National Academy of Sciences, Engineering and Medicine (NASEM) study committee on Future Biotechnology Products and Opportunities to Enhance Capabilities of the Biotechnology Regulatory Systems, as it is a subject about which I've written extensively. That committee's work is underway at the time of the writing of this chapter, and I find myself now on the other side of the table at the National Academies. As a society, we are at a crucial time in biotechnology policy. New genetic engineering technologies like gene editing, gene drive systems, synthetic biology, and de-extinction are challenging our abilities to keep pace from a governance perspective. On top of biotechnologies are emerging capabilities in nanotechnology, robotics, neuroscience and neurotechnology, information technology, data sciences, and geo-technology that are converging with genetic engineering.

I truly hope that we can come together in a democratic society to open up the decision making processes to a wider range of social science, cultural, ethical, and demographic perspectives. We are in need of science-informed, value attentive governance systems to envision and guide technology down paths that most benefit society and lead to human happiness and health and ecological sustainability. We need to hear from many voices, including youth, to envision the society we would like our children and grandchildren to inherit. In the context of GEOs, will it be the market, technological elites, or many publics who will decide whether and which GEOs are deployed into products and the environment? I personally hope for the latter.

Most recently, I have become very interested in the gender, intergenerational (Kuzma and Rawls 2016), and cultural equity issues surrounding technological decision making. For example, In the social science literature (under the rubric of cultural theory), it has been found across multiple studies and science and technology areas that women and under-represented racial or ethnic groups have higher concern about technological risks and the environment, while white males have a lower level of concern, even when education, income, and other demographic factors are accounted for (Kahan 2012). Females and minorities are also more likely to have egalitarian and communitarian political leanings, as opposed to hierarchical and individualistic ones held by white or Caucasian males (Kahan et al. 2007). Yet, leaders and decision-makers (e.g. division directors in government or company executives whom interact with them) are disproportionately Caucasian males. As a woman relatively advanced in her career, I am often the only or one of very few females at higher-level meetings, panel discussions, or workshops. The current decision making system seems to me to be unjust. An opening up of regulatory processes to a greater diversity of people and perspectives might remedy this inequity and increase procedural justice. So now, I am returning to that social justice interest I had as an undergraduate again. The beauty of policy sciences is the ability to learn and return to different, but related topical areas.

I come with my own biases to the agricultural biotechnology debates, as I've freely expressed in this chapter. We all have these biases and must be cognizant of them. We also must respect alternative perspectives and biases that do not match our own. As I traveled on my career path, these biases were shaped by my background in ethics and philosophy, biochemistry and molecular biology, risk analysis, science and technology policy, and the social sciences. I hope to see the current biotechnology revolution shaped by many different viewpoints so it is done in the best interest of all of society, not just a few groups (like technology-elites or white-males in the United States). Only then will we be able to move past the inflamed and divisive rhetoric and enable safe, responsible, socially desirable and appropriate use of genetic engineering.

References

Bardocz S, Clark A, Ewen S, Hansen M et al (2012) Seralini and science: an open letter.' Independent Science News [Online], 2 October. http://independentsciencenews.org/health/seralini-and-science-nk603-rat-studyroundup/

Bøhn T, Cuhra M, Traavik T, Sanden M, Fagan J, Primicerio R (2014) Compositional differences in soybeans on the market: Glyphosate accumulates in Roundup Ready GM soybeans. Food Chem 153:207–215

Carman JA, Vlieger HR, Ver Steeg LJ, Sneller VE, Robinson GW, Clinch-Jones CA, Edwards JW et al (2013) A long-term toxicology study on pigs fed a combined genetically modified (GM) soy and GM maize diet. J Org Syst 8(1):38–54

Domingo JL (2016) Safety assessment of GM plants: an updated review of the scientific literature. Food Chem Toxicol 95:12–18

Domingo JL, Bordonaba JG (2011) A literature review on the safety assessment of genetically modified plants. Environ Int 37(4):734–742

Dona A, Arvanitoyannis IS (2009) Health risks of genetically modified foods. Crit Rev Food Sci Nutr 49(2):164–175

Ebel E, Schlosser W, Kause J, Orloski K, Roberts T, Narrod C, Powell M et al (2004) Draft risk assessment of the public health impact of Escherichia coli O157: H7 in ground beef. J Food Prot 67(9):1991–1999

EPA (1994, 2001) Environmental Protection Agency. Regulations under the Federal Fungicide, Insecticide, and Rodenticide Act for plant-incorporated protectants. Fed Reg 66:37855–37869

FAO (2016) United Nations Food and Agricultural Organization. Agriculture and Food Security. http://www.fao.org/docrep/x0262e/x0262e05.htm. Accessed 14 July 2016

Glöckner G, Séralini GE (2016) Pathology reports on the first cows fed with Bt176 maize (1997–2002). Sch J Agric Sci 6:1–8

Gu J, Bakke AM, Valen EC, Lein I, Krogdahl Å (2014) Bt-maize (MON810) and non-GM soybean meal in diets for Atlantic Salmon (Salmo salar L.) juveniles-impact on survival, growth performance, development, digestive function, and transcriptional expression of intestinal immune and stress responses. PLoS ONE 9(6):e99932

Hilbeck A, Binimelis R, Defarge N, Steinbrecher R, Székács A, Wickson F, Novotny E et al (2015) No scientific consensus on GMO safety. Environ Sci Eur 27(1):1

Kahan DM (2012) Cultural cognition as a conception of the cultural theory of risk. Handbook of risk theory. Springer, New York, pp 725–759

Kahan DM, Braman D, Gastil J, Slovic P, Mertz C (2007) Culture and identity-protective cognition: explaining the white-male effect in risk perception. J Empirical Legal Stud 4 (3):465–505

Kuzma J (2013) Properly paced? Examining the past and present governance of GMOs in the United States. In: Marchant GE, Abbott KW, Allenby B (eds), Innovative governance models for emerging technologies, Edward Elgar, Cheltenham, UK, pp 176–197

Kuzma J (2016a) Reboot the debate on genetic engineering. Nature 531:165–167

Kuzma J (2016b) Risk, Environmental governance, and emerging biotechnology. In: Durant R, Fiorino DJ, O'Leary R (eds) Environmental governance reconsidered: challenges, choices, and opportunities, 2nd edn. MIT Press, Cambridge

Kuzma J, Besley JC (2008) Ethics of risk analysis and regulatory review: from bio- to nanotechnology. Nanoethics 2(2):149–162

Kuzma J, Rawls L (2016) Engineering the wild: gene drives and intergenerational equity. Jurimetrics J Law Sci Technol 56(3):279–296

Kuzma J, Najmaie P, Larson J (2009) Evaluating oversight systems for emerging technologies: a case study of genetically engineered organisms. J Law Med Ethics 37(4):546–586

Leiss W, Powell D (1997) Mad cows and mothers milk. McGill–Queen's Press, Montreal

Loening UE (2015) A challenge to scientific integrity: a critique of the critics of the GMO rat study conducted by Gilles-Eric Séralini et al. (2012). Environ Sci Eur 27(1):1

Losey J, Raynor L, Carter M (1999) Transgenic pollen harm monarch larvae. Nature 399:214

Lurquin PF (2016) Production of a toxic metabolite in 2, 4-D-resistant GM crop plants. 3 Biotech 6(1):1–4

Meyer H, Hilbeck A (2013) Rat feeding studies with genetically modified maize—a comparative evaluation of applied methods and risk assessment standards. Environ Sci Europe 25:1–33

NAS (1987) National Academy of Science. introduction of recombinant DNA-engineered organisms into the environment: key issues. National Academy Press, Washington

NASEM (2016) National Academies of Science, Engineering and Medicine. Genetically engineered crops: experiences and prospects. National Academy Press, Washington

NRC (2000) National Research Council, *genetically modified pest-protected plants: science and regulation*. National Academy Press, Washington

Office of Science and Technology Policy (1986) Coordinated framework for the regulation of biotechnology. Fed Reg 51(123):23302–23350

USDA (1993) US Department of Agriculture, genetically engineered organisms and products: notification procedures for the introduction of certain regulated articles and petition for nonregulated status. Fed Reg 58(60):17044–17059

USDA (1997) Introduction of organisms and products altered or produced through genetic engineering which are plant pests or which there is reason to believe are plant pests, 7 C.F.R. § 340

Vecchio L, Cisterna B, Malatesta M, Martin TE, Biggiogera M (2009) Ultrastructural analysis of testes from mice fed on genetically modified soybean. Eur J Histochem 48(4):449–454

Enabling Educators: Biotechnology in the Classroom

Kathleen E. Kennedy

Preface

The North Carolina Biotechnology Center (hereinafter referred to as "the Center" or "NCBiotech") is an independent, non-profit corporation largely funded by the State of North Carolina to stimulate the growth of biotechnology industry in the state. I became acquainted with the Center when I was on the faculty at East Carolina University in Greenville, North Carolina. My colleague Edmund Stellwag was teaching biotechnology to high school teachers in summer workshops sponsored by the Center. Ed is a gifted teacher, and I could see that both he and his audience were having an intense and wonderful time. (He kept teaching teachers for many years.) I later heard that the Center's Education and Training Program was recruiting for a new position. The staff wanted to expand their scope of operations beyond secondary schools to colleges and universities, and it seemed I had suitable qualifications: teaching experience, a bachelor's degree in Journalism and MS in Botany from the University of Texas at Austin, a Ph.D. in Molecular Biology from Vanderbilt University, and postdoctoral research at the Biozentrum of the University of Basel in Switzerland.

I joined the Center in 1990 as Workforce Training Manager and found it an interesting place to work with able and congenial colleagues in a beautiful setting. I liked the variety of projects and I liked working on the big canvas of a whole state. In my view, the hallmarks of the Center always have been a generous hospitality, thoughtful and effective program development, and high quality standards. In time I became the Vice President of the Education and Training Program. I left in 2013 with the rest of my staff when the state legislature cut the Center's funding by 30%, forcing reductions throughout the organization.

K.E. Kennedy (✉)
Education and Training Program, North Carolina Biotechnology Center, Durham, USA
e-mail: kkennedync@earthlink.net

© Springer International Publishing AG 2017
L.S. Privalle (ed.), *Women in Sustainable Agriculture and Food Biotechnology*,
Women in Engineering and Science, DOI 10.1007/978-3-319-52201-2_7

Acknowledgements

I am most grateful to colleagues in the Education and Training Program during my tenure at the Center whose hard work and dedication made possible all the accomplishments described herein: John Balchunas, Lynn Elwell, Ph.D., Helen Kreuzer, Ph.D., Rob Matheson, Maria Rapoza, Ph.D., and William (Bill) Schy, Ph.D.; to our outstanding administrative assistants, Amy Black and Jessica Willsey; to the able contractors who supported our projects: Jean Chappell, Catherine Dieck, Ph.D., Julie Omohundro, Carol Schafer, Brenda Summers; and the *BioWork* team— Joseph Delmonico, Pat Hill, and Sandy Thomas. In addition, I wish to thank all the Center staff in other divisions who helped us along the way for many years and who continued to help me with information for this article.

It was a great pleasure to have worked with Vice President Adrianne Massey, Ph.D. and Senior Vice President Ken Tindall, Ph.D. who were not only my supervisors, but also valued mentors and friends from whom I have learned so much.

I also wish to express my thanks to the many professionals throughout North Carolina who were essential contributors to our work:

- The instructors who taught summer workshops and the many teachers who implemented what they learned in their classrooms;
- The industry employees who opened their doors to us and educated us about their work;
- The advisors who reviewed education grant proposals and helped shape our funding philosophy, and the faculty whose new initiatives we funded.

Making the Connections that Make Business Happen

The focus of this chapter is the Education and Training Program at the North Carolina Biotechnology Center and how it addressed public education about biotechnology as well as education to build the workforce in support of industry growth. Since the work of this unit took place within the context of the Center's mission as a whole, an overview of the Center's operations may be helpful. (For a full description of all the Center's activities, see www.ncbiotech.org) (Figs. 1 and 2).

The Center has served since 1984 as a focal point for North Carolina's steadily growing biotechnology community. Over time, its mission has expanded to stimulate broader life-science based economic development in North Carolina by supporting innovation, commercialization of new technologies, and business growth. The Center's resources for serving the life science community are:

- Expert staff with knowledge and experience in science and business;
- Funding for research, small business start-up, and regional economic development to support projects at early stages when there are few other sources of financial assistance;

Fig. 1 The Biotechnology Center's headquarters site in the Research Triangle Park houses staff offices as well as a conference center with meeting rooms of different sizes, a 170-seat auditorium, and open space to accommodate vendor shows or poster sessions. Additional office space can be rented by start-up companies (*Source* NCBiotech/Jim Sink)

- Library research services and extensive industry-relevant databases;
- Five regional offices across the state in addition to the Research Triangle headquarters (Fig. 2).

Communication is a major role for the Center both to the public and to specific audiences about the science and business of biotechnology, as well as the Center's programs. The Center promotes North Carolina worldwide as an ideal place to locate life science enterprises. Making connections with the right people is essential to the success of researchers, entrepreneurs, and investors; and Center staff, with their wide knowledge of the biotech community, facilitate these connections. Since the Center is neither a university nor a business nor an agency of the state government, it can act as a neutral facilitator and convener for all these sectors, and partners with all of them in building new initiatives.

Examples of such partnerships include:

- NCBioImpact—a partnership of the Center with industry, universities and community colleges to build facilities and academic programs for workforce training; see http://ncbioimpact.org.

Fig. 2 NCBiotech interior, 2nd floor hall leading to open conference space. To the *left* the 1st floor reception area can be seen through the railing

- Collaborative Funding Grants—funded jointly by the Center and the Kenan Institute for Engineering, Technology and Science (https://kenan.ncsu.edu/start. html) to support university-business research projects. See http://www. ncbiotech.org/research-grants/research-funding/collaborative-funding.

Bang for the Buck

NCBiotech Library data show that as of June 30, 2015, the Center had provided over $168 million in low-interest loans to startup companies and grants to advance life science research, education, and other initiatives. Work supported by these funds has enabled recipients to obtain additional funding from other sources. Every dollar awarded by the Center in grants to universities has led to follow-on funding for the recipients in the range of $4–$88 (depending on the grant program). Every dollar awarded in loans to small companies has led to an average of $112 in follow-on funding. For agriculture-related business loans, the average follow-on funding figure was $305. Small company loan recipients operating in 2014 generated estimated revenues of $1.9 billion (Battelle Technology Partnership Practice 2014a).

The Center's operations, complementing North Carolina's great universities, the growing intellectual capital in the Research Triangle region, quality of life and advantageous location have helped to make North Carolina one of the largest concentrations of biotechnology industry in the country since 2004 (Ernst and Young 2004). As of 2015, the Center had identified more than 600 life science companies with an estimated 63,000 employees that generated over $73 billion in total economic activity in the state in 2014 (Battelle Technology Partnership Practice 2014b). Between 2001 and 2012 North Carolina life science employment grew by 30.0% compared to 7.4% for the US as a whole (Battelle Technology Partnership Practice 2014c).

The value of the Center's role during 32 years of building connections and new initiatives is reflected in this quotation from Michael Porter: "...the enduring competitive advantages in a global economy lie increasingly in local things—knowledge, relationships, motivation—that distant rivals cannot match" (Battelle Technology Partnership Practice 2014d).

Agricultural Biotechnology in North Carolina

In the early 1980s to mid-1990s major agricultural chemical-producing companies saw the potential of recombinant DNA technology and wanted to benefit from the research at universities and biotech companies in the Research Triangle area. In addition, the state had a significant agricultural business and production sector. North Carolina has one of the most diversified agricultural outputs in the nation, growing over 80 different commodities (Shank and Niebauer 2016).

In an interview in January, 2016, Scott Johnson, Vice President of Agricultural Biotechnology at the Center, who has long experience in the industry, recalled this period in North Carolina as one of "scientific ferment," enhanced by the presence of Dr. Mary Dell Chilton and resulting in a continuous expansion of the ag biotech business community. Now, he also sees more growth in capacity building by established companies (e.g., new greenhouses and laboratories) for commercial development of new products. Scott stated that there is no other ag biotech cluster in the world that has North Carolina's combination of intellectual capital, a thriving business community, and trained workforce; and that these are the factors that continue to draw more companies to locate here.

Growth of the Industry

The Research Triangle area is now home to major research and development facilities of several of the largest global ag biotech companies: BASF Plant Science, Bayer CropScience, DuPont Pioneer, Syngenta, and (until 2016) Monsanto; as well as many smaller companies. Total employment in ag biotech is over 8,900—an increase of 20.3% since 2010. These companies are involved in diverse areas: plant and animal health and productivity, environmental remediation, industrial compounds and pharmaceuticals (Shank and Niebauer 2016).

In 2010, the Biotechnology Center launched a new organizational unit focused on agricultural biotechnology, now headed by Scott Johnson. The Agricultural Biotechnology Initiative holds quarterly, annual and biennial events for the ag biotech community, including an Entrepreneurial Showcase to connect entrepreneurs with investors (http://www.ncbiotech.org/content/signature-programs). The unit also maintains an online database of North Carolina ag biotech companies that are seeking funding, partners or licensing opportunities (http://www.ncbiotech.org/agBIOCEPS).

New Chow for Hogs and Chickens?

North Carolina is the second largest producer of swine and poultry in the nation, and they eat a *lot*—three times more than the 100 million bushels of corn and other feed grown in the state each year. And it's a problem in other mid-Atlantic states as well. So producers have to import grain from the Midwest at higher feed prices. A new initiative by the Biotechnology Center —the Crop Commercialization Center (CCC)—could change this situation. The CCC, a collaborative venture of universities, growers, and other industry groups in Mid-Atlantic states is headed by Paul Ulanch, Ph.D., MBA. The CCC's current goal is to find ways to produce at least 50% more feed grain. This region is not ideal corn-growing country, so other alternatives are needed, and grain sorghum (milo) is the object of current field trials carried out by North Carolina State, Clemson, Virginia Tech and the University of Maryland to test different varieties. Sorghum typically thrives in hot, dry conditions and the challenge in the Mid-Atlantic will be higher moisture levels and the need to develop fungal-resistant varieties. (http://www.ncbiotech.org/article/regional-consortium-builds-livestock-feed-grain-system/67216; http://www.ncbiotech.org/content/bccc).

The Plant Molecular Biology Consortium

The Plant Molecular Biology Consortium (PMB) was an early facilitator of the "scientific ferment" Scott described. In 1984–85 Duke University, North Carolina State University, North Carolina Central University and the University of North Carolina at Chapel Hill joined with the Biotechnology Center to form a consortium to advance research in plant molecular biology. CIBA-GEIGY (now Syngenta), R. J. Reynolds, and the Center contributed funding for 2 years to support graduate and post-doctoral fellowships. These were intended to recruit outstanding candidates from around the world.

Dr. Laura Privalle was the recipient of one of these fellowships. As she recalls, "A PMB fellowship in 1984 set me on the path of my future career. I was a post-doc in Dr. Henry Kamin's lab in the Biochemistry Department at Duke University when I applied for one of these fellowships. Someone from CIBA-GEIGY was on the panel that reviewed the applications and asked me to apply there; and I was hired as a postdoctoral fellow, working in plant protein biochemistry." For the rest of Laura's story, see [chap. "The Story Behind the Approval of the First *Bt* Maize Product"]

The PMB continues to be self-supporting through sponsorships from agricultural biotechnology companies. In addition to regular seminars throughout the year by internationally recognized speakers, the PMB also has annual weekend retreats, held in the mountains or on the coast of North Carolina in alternate years. http://www.ncbiotech.org/business-commercialization/connect-with-colleagues/intellectual-exchange-groups/PMB. The Center has assisted in the formation of other scientific Exchange Groups focused on a variety of topics in the Research Triangle area and across the state. http://www.ncbiotech.org/exchange-groups.

Education and Training Program

The Education and Training Program (ETP) worked to increase public understanding and acceptance of the applications of biotechnology, and encourage the development of a well-educated workforce to support industry growth. Outreach to the public was a particular interest of Adrianne Massey, Ph.D., who was head of the Program when I started work at the Center in late 1990. Adrianne now is Managing Director, Science and Regulatory Affairs, Food and Agriculture, Biotechnology Innovation Organization (BIO). Adrianne initiated the Center's Secondary Education Program to reach the public through the education system by enabling middle and high school teachers across the state to bring biotechnology into their classrooms. This program is described in detail below, as are our other activities in researching and addressing workforce development needs and providing grants for biotechnology education projects.

ETP staff also worked to make the connections that make good science education happen by:

- Producing publications about the science and applications of biotechnology both for classroom use and to inform educators about preparing graduates for jobs; and
- Holding conferences for educators and visiting schools, colleges, and universities across the state to advise faculty and provide information about biotechnology science, industry, careers, and curriculum design.

Teacher Empowerment

"Teachers need support in so many ways…. Many teacher training programs do not provide the instruction necessary for teachers to feel comfortable setting up biotechnology labs and activities. NCBC's program empowers teachers, gives them the opportunity to learn, and provides the resources so they can implement what they have learned. NCBC was essential to the success of my students!"—2010 workshop participant (NCBiotech survey data).

Professional Development for Secondary School Teachers

People who don't attend college usually get their only formal science education in middle school and high school, and nurturing an interest in science needs to happen at an early age. Interests that lead to future career choices often are formed in the middle school years. A child's excitement also can motivate her parents to learn more about whatever fascinates her.

Hence, introducing biotechnology into school curricula has strategic value. But many teachers get a general science education in college and may have little or no exposure to biotechnology. Even many who have a BS degree in biology—especially older teachers—may have had limited exposure to biotechnology or related fields. Educating teachers about the science and applications of biotechnology was therefore essential. This approach was not unique to North Carolina—several other states have also provided ongoing teacher training in biotechnology.

By 2013, about 2,000 teachers from 99 of North Carolina's 100 counties had attended Center-organized professional development workshops. Each teacher typically instructs on average about 100 students a year—so the multiplier effect over time is significant. I estimate that a typical graduating class of teachers in a summer in which we offered six workshops could reach 9–10,000 students in the following school year, and for many years thereafter.

Summer Workshops for Teachers

In developing the Secondary Education Program Adrianne first set up a board with representatives from industry, academia, and the public, as well as a teacher advisory board and a scientific review panel. She worked with the state Department of Public Instruction (DPI), which selected a group of 28 motivated and capable "master teachers" to attend the initial workshop in 1987, taught by Dave Micklos of Cold Spring Harbor Laboratory. The master teachers developed lesson plans, field tested lessons, materials, and lab equipment, and served as co-instructors with college faculty at subsequent workshops. DPI also provided teacher license renewal credits, helped with workshop coordination and teacher selection for other early workshops, and provided funding for initial lesson plan development.

ETP staff whose primary responsibilities included day-to-day management of this program were (in order of succession over 26 years): Dave Smith, Rob Matheson, Helen Kreuzer, Ph.D., Lynn Elwell, Ph.D., and Bill Schy, Ph.D. Typically they ran five to eight workshops each summer, located at college and university campuses across the state. Most workshops ran five days. Participants were provided with free room and board and a $50 *per diem*. Each workshop was co-taught by a university or college instructor and a master teacher. Over time, a group of engaging and capable instructors were recruited, several of whom taught workshops year after year. Teachers are a wonderfully appreciative audience; and the instructors, despite the hard work required to put on a workshop, found it a rewarding experience.

Dr. Maria Rapoza, Workforce Training Manager from 1999–2002, was familiar with the Center's workshops from her previous work at Carolina Biological Supply (an educational laboratory supply company). She recalls being impressed by the engagement and enthusiasm of teachers she worked with when she visited Center workshops. She said in an interview "When the Center started in the 80's, biotech was like outer space: cool, cutting edge, but not part of everyone's daily life;" and observed that because our studies of the industry and its employment needs informed our teacher education projects, teachers could see biotechnology not just as "rocket science" but as a real presence in North Carolina with useful products and job opportunities for students with a range of abilities and interests. Maria thinks this broader message enabled teachers to more readily gain interest and support from school administrators and parents for implementation of biotech content in their schools. (Later, after a brief stint in industry, Maria came back to the Center as head of the Science and Technology Development Unit until 2015.)

Workshop Content

A typical slate of five to eight workshops in any given year might have included at least two introductory high school level workshops in general biotechnology (with a prescribed standard content); a middle school workshop; and two or more "Special Topic" workshops in subjects such as agricultural biotechnology, marine biotechnology or stem cell technology. Many teachers attended multiple workshops in different topics over the years.

Formal lectures by the college faculty were balanced with hands-on time for the teachers to practice grade-appropriate classroom activities and lab experiments. The master teachers led these activities, discussed pedagogy and lesson plans, and showed teachers how topics were linked to the state-mandated course of study.

Activities included "dry labs" which can be done with everyday items in an ordinary classroom without sinks or lab equipment.

Examples of Introductory High School Workshop Activities
Materials kits for many classroom activities including some listed below are available from Carolina Biological Supply in Burlington, NC.

- Fruit DNA extraction (frozen strawberries work well) with kitchen detergent and alcohol
- Gel electrophoresis of lambda phage and/or plasmid DNA restriction digests
- Transformation of *E. coli* with plasmid DNA
- Construction paper modelling of recombinant plasmids
- Polymerase chain reaction—actual or modeled with chains of colored paper clips
- Microbial degradation of oil
- Yeast growth in varying culture conditions
- Microarray simulations (on paper).

Publications for Teachers

Adrianne and Helen authored *Molecular Biology and Biotechnology: A Guide for Teachers*, which was given to teachers as a reference book in most of the workshops (Kreuzer and Massey 2008). This text evolved in part from earlier workshop manuals. Helen and Adrianne later wrote a second textbook intended for college freshman level introductory or non-majors biology classes: *Biology and Biotechnology: Science, Applications, and Issues* (Kreuzer and Massey 2005). This textbook includes content on the uses of technology and controversial issues arising from them, and was provided in workshops on agricultural biotechnology and stem cell research.

Career Pathways: Focus on Biotechnology was produced at the request of the state Department of Public Instruction (Balchunas and Omohundro 2006). John Balchunas, Workforce Training Manager (later Director) was the lead author with Julie Omohundro and I was editor. This booklet was a guide to jobs in the industry and what kinds of education students would need to prepare for them. It was written at student level and was our most popular publication for teachers. Every workshop attendee received a classroom set of 25 copies and we circulated them widely in other venues.

Heal, Feed, Sustain: How Biotechnology Can Help Save the World, is a video filmed on location featuring middle-school students visiting the Institute for

Regenerative Medicine, (www.wakehealth.edu/WFIRM), Novozymes North America, and Syngenta. This video and an accompanying Teacher's Guide were produced by the North Carolina Association for Biomedical Research (NCABR) with advice and funding from the Center. These materials can be downloaded from: http://www.ncabr.org/k12/biotechshortfilm.

Earlier teacher resources (no longer publically available) were the *Carolina Genes* newsletter that provided articles on biotech topics and lab activities; and *Biotechnology: Sowing the Seeds for Better Agriculture*, a video about plant genetic engineering and its potential for crop improvement produced with the collaboration and support of Syngenta.

Follow-up Support

The impact of workshops would have been limited if teachers were unable to implement what they learned because schools lacked lab equipment and supplies. We addressed this problem in two ways. First, we included in almost all workshops activities that could be done with ordinary home or office items, as well as suggestions for class discussions or web exploration. Second, we provided lab supplies and equipment. Every workshop graduate received an annual free lab supplies order form listing all the items needed for experiments taught in all workshops. Teachers could select items up to a value of $160–$200 (depending on our budget), and send the order to Carolina Biological Supply. The Center paid the bill. Trunks loaded with gel boxes, power supplies, microfuges, micropipettors, and thermal cyclers were shipped on loan to schools.

Even so, usage of these resources was limited. Only about 20–30 teachers took advantage of the loaner equipment in any given year. Teacher feedback indicated that many didn't have time in their classes for activities requiring the equipment (50-min periods are too limited), or still felt a lack of confidence in using it, or their teaching assignments had changed. Some teachers indicated their schools had been able to buy equipment, sometimes with Center grant funds. Many more teachers ordered supplies. In the 2010–11 school year, for example, 270 teachers placed orders.

Workshop Outcomes

We routinely obtained feedback from attendees at the end of every workshop with a simple evaluation form, but it also was essential to gauge the extent to which teachers were actually using what they had learned in their classrooms. Adrianne sent out a survey in 1996 to 550 workshop graduates from the previous 9 years; 134 were returned. The results are listed in Table 1.

Table 1 Results of 1996 survey of graduates from workshops held from 1987 to 1996. The right-hand column lists percentages of teachers reporting that they had taken actions or agreed with opinions in the statements listed in the left-hand column

Added information about biotech to their courses	100%
Are more confident about their knowledge of cellular/molecular biology	98%
Think they are better teachers	98%
Have expanded or significantly changed an existing course	74%
Have created a new biotechnology course	15%
Have encouraged other teachers to attend a center workshop	91%
Have influenced other teachers to add biotechnology to their courses	76%
Cite evidence that their students are better informed about biotechnology	90%
Think knowledge has improved student attitudes about biotechnology	93%
Think knowledge has sparked some student career interest in biotechnology	67%

Eighty-nine teachers (67%) responded to an invitation to provide comments. All but two responders provided comments expressing praise and gratitude for the program, excitement about their classroom implementation, how much the supplies and equipment were needed, or requests for more and/or different workshops. A few examples:

- "It has been the most helpful, productive science education project that I have ever participated in."
- "You have revolutionized my teaching through workshops."
- "Students are not afraid of 'genetic monsters' now."
- "I encouraged a fellow teacher to attend [a workshop] last year, he was so excited about what he had learned. He now incorporates many of your activities in his teaching."

Workshop Evaluation: 2010–2012

Staff organized 23 workshops during this 3-year period at ten institutions across the state, with a total enrollment of 419. The standard five-day introductory biotechnology workshop for high school teachers was offered eight times. Eleven other workshops included bioprocessing and the ever-popular "Microbial Magic" for middle school teachers; and workshops for high school teachers on agricultural biotechnology, marine biotechnology, and stem cell technology. Some of these were offered more than once.

In 2010 I engaged a professional education evaluator, Eleanor Hasse, Ph.D., to devise new instruments for a more comprehensive evaluation. These included questionnaires administered at the beginning and end of each workshop and observation visits by Eleanor and her colleagues. The results across all workshops in 2010, 2011, and 2012 are described below.

Survey Responses. Well over 90% of the participants rated the general atmosphere of the workshops as very helpful and friendly, the benefit they had derived greater than they had expected, and the quality of instruction very good to excellent. The *level* of instruction, as compared to *quality* of instruction was rated lower. Average annual scores for a level of "just right" ranged from 67 to 86%. I think this may be attributable to differences in prior background knowledge of teachers; as well as to a few inherently difficult workshop topics and a few new instructors introduced during this period.

Overall, participants acquired a major increase in comfort level with lab safety procedures as a result of their workshop experience. Almost all teachers identified at least a few workshop activities that they definitely planned to use in their classrooms. They were more likely to choose simpler activities that do not require specialized equipment, even though the Center would have provided it.

Asked about barriers that would be likely to prevent implementation of what they had learned, teachers in the 2010 workshops cited one or more of the following issues:

- Lack of instructional time
- Lack of time for planning and set-up
- Emphasis on end-of-course tests
- Classes too large
- No appropriate lab space
- Learning objectives don't align with those for courses taught by some teachers.

This list underscores the importance of providing teachers with quick, simple activities which can address both time and class size constraints and lack of lab space.

All surveys included an invitation to make open-ended comments and suggestions, and the great majority of these were positive and useful. Teachers most frequently identified cost-effective hands-on lab activities, Center resources, learning real-world applications of biotechnology, and general knowledge of the subject as the most important things learned. Many teachers wanted more workshops on different topics. Negative comments were rare.

Some examples:

- "Workshop was one of the best I have attended in 11 years teaching. I have attended very few science-specific workshops that have taught me so much in a limited amount of time."
- "I…learned what employers in biotech companies find most valuable in employees which will directly affect how I structure my hands-on labs as well."
- "Biotechnology is not as scary as I thought it was. There are so many simple things that I can use in my classroom and that will make a huge difference."

Follow-up Survey. In December 2010, Eleanor emailed a short electronic survey to all 141 participants in the summer 2010 workshops and received 60 responses.

We were pleased to see that 77% of the respondents had used workshop activities in their fall classes. Most of the respondents who didn't had not yet taught a relevant course or unit.

The most popular activities were class discussions, dry labs, and DNA extraction, followed by gel electrophoresis, growing bacteria and yeast, and enzyme experiments. On average, 90% of respondents who had introduced workshop activities into their classes indicated that the activities helped students understand concepts, generated student interest in biotechnology, were easy to implement, and worked as planned. Teachers also had used Center resources, including free lab supplies, the Center's website, and the *Career Pathways* publication.

Instructor Surveys. Eleanor also surveyed all the workshop instructors and master teachers who taught in the eight 2010 workshops. All of the respondents were enthusiastic about the workshops and eager to continue to participate. They found teaching workshops beneficial both personally and professionally and expressed great satisfaction in helping K-12 teachers and students. They identified a wide array of benefits to the teacher participants, the most frequently mentioned being the opportunity to collaborate with others, increased confidence with laboratory work, increased conceptual understanding, and increased enthusiasm.

Quotes from instructors:

"I received an entire new perspective on teaching this material. I learned all types of new ideas that I intend to modify and utilize in my introductory cell/molecular biology courses. I gained a great deal of respect for the teaching that occurs in the high schools, particularly when budgets are a small fraction of what we receive at my institution."

"I feel instructing these workshops is the single most direct way in which I impact the scientific literacy of young students, better preparing them to be informed voters and maybe even piquing the interest of future biologists."

What do you hope to learn from this workshop?

"I was raised on a farm and have seen so many changes over the past 45 years. [When I was] a child my family worked peanuts with a hoe for weeks on end. There was no cotton grown in northeastern North Carolina in the late 60's—the early 80's but now it is king again. My brother farms about 4,000 acres and I often ask him about seed prices, herbicide and pesticide applications, and the role genetic engineering plays in his operation. I feel so fortunate to have the opportunity to attend such a prestigious workshop for such a nominal fee. I look forward to sharing the knowledge I gain with my students and my brother." [*Pre-workshop survey question response from a teacher enrolled in a 2015 agricultural biotechnology workshop*].

Agricultural Biotechnology Workshops for Teachers

The general biotechnology workshops usually included some coverage of genetic modification of plants; and there had been three special topic workshops on agricultural biotechnology. But the Center's new Agricultural Biotechnology Initiative and the growing concentration of ag biotech companies in our neighborhood led me to the idea for a new kind of workshop with the collaboration and support of industry. Syngenta, BASF, Bayer CropScience, and Monsanto liked the idea and agreed to provide financial support and employee assistance. (These were not the only companies interested in public understanding of biotechnology—Biogen has Community Labs at both its Research Triangle and Cambridge, MA sites that offer summer sessions for students.)

Bill and I worked with a group of great educators to develop the detailed content and agenda for the first workshop in 2012 for middle school teachers: Karthik Agorham, Ph.D., Professor of Biology at Meredith College in Raleigh, Ms. Cinnamon Frame and Ms. Lori Stroud, experienced master teachers in science and agriculture, respectively, and Deborah Thompson, Ph.D., who worked in the Center's Science and Technology Development group and had a background in ag biotech.

The prototype middle school workshop (2012) at North Carolina Central University was taught by Deborah as lecturer, with Cinnamon and Lori as master teachers. As these workshops were more complex than usual, I also hired Catherine Dieck, Ph.D. to take care of logistics. Karthik taught the prototype workshop for high school teachers in 2013, with Jon Davis, a science instructor at the North Carolina School of Science and Mathematics who has developed a 4-year high school agriculture curriculum, and Cathy Berrier, agriculture teacher. University and industry scientists were invited as guest lecturers (Fig. 3).

These 5-day workshops are similar to our other workshops, but have the following unique features particularly appreciated by the teacher participants:

- Company sponsors provide facility tours and opportunities to visit with employees.
- "The View from the Field" is a 2 h panel discussion featuring organic and conventional growers.
- An industry scientist lectures on testing the safety of genetically modified crops for consumers and the environment.
- A "Tech Expo" period provides an exhibition of simple experiments, websites, and teaching materials during which teachers can browse, share ideas, and consult with instructors.
- Workshop content is designed to benefit both science and agriculture teachers.
- The text is Tomorrow's Table: Organic Farming, Genetics, and the Future of Food, by Ronald and Adamchak (2010).

Since the closure of the Education and Training Unit, the Center's Agricultural Biotechnology group has continued to organize these workshops for both middle

Fig. 3 Teachers working on an experiment at an agricultural biotechnology workshop

and high school teachers with support from industry. Workshop information and detailed agendas for current workshops are available on the Center's website at: http://www.ncbiotech.org/agbiotechclassroom.

Outcome Evaluation

Outcomes overall have been very good, similar to those of our other workshops; and indicate an enthusiastic response by participants. Teachers have valued especially the opportunities to converse with company employees and growers. Eleanor noted "The results show overall a group of teachers already knowledgeable and comfortable growing more so because of the workshop." This may reflect to some degree the presence of agriculture teachers in the group. A feature in the evaluation process for these workshops is pre- and post-workshop questionnaires that assess different aspects of teacher understanding of the subject and their general comfort level in teaching about it (See Table 2 for an example).

The middle school teachers were somewhat less knowledgeable coming into the workshops than the high school teachers, but by the end of the workshops about 95% of participants in both groups agreed or strongly agreed that they were

Table 2 Consolidated pre- and post-workshop participant responses to the statement: "I am comfortable teaching about controversial ethical, social, environmental or economic issues related to applications of biotechnology in agriculture." Data are from three middle school and three high school ag biotech workshops. Participants were asked to rate their responses as: 1 = Strongly disagree; 2 = Disagree; 3 = Neutral or not sure; 4 = Agree; 5 = Strongly agree

Rating	Number of responses			
	Pre-workshop		Post-workshop	
	Count	%	Count	%
Middle school workshops				
1	1	2.3	0	0
2	9	20.5	0	0
3	10	22.7	2	4.8
4	18	40.9	25	59.5
5	6	13.6	15	35.7
Total	44	100	42	100
High school workshops				
1	0	0.0	1	1.9
2	3	5.7	0	0.0
3	18	34.0	2	3.8
4	25	47.2	23	43.4
5	7	13.2	27	50.9
Total	53	100	53	100

comfortable teaching about controversial issues related to ag biotech. The greatest gain in knowledge was seen in response to the statement: "I understand how GMO (genetically modified organism) food safety is evaluated." The fraction of teachers selecting ratings of 4 or 5 increased from only 21.2% pre-workshop to 89.9% by the end of the workshop.

Post-workshop surveys in 2013 also requested responses from teachers to the statement: "Many of my students oppose GM crops." Fifty-percent of the 34 teachers in the middle and high school workshops in 2013 responded as neutral or not sure (rating 3). Only 2 of 17 middle school teachers agreed with the statement but 6 of the 17 high school teachers did. This is a small amount of data, but possibly suggests that younger students might be a more open-minded audience.

Public Opinions About Scientific Issues

At a workshop convened by the National Academy of Science on public opinion about GMOs, "Public Engagement on Genetically Modified Organisms: When Science and Citizens Connect," (National Academy of Sciences 2015). Dr. Dominique Brossard of the University of Wisconsin-Madison presented data from public-opinion polls over the last decade. She observed that there are "very vocal minorities" [about 10% pro and con] that hold strong views; but that "roughly one-third of the American populace does not care—does not have sufficient time, energy, or interest to invest intellectual energy in the GMO debate." (According to the workshop outcomes reported above, children might be less polarized than their parents.)

This National Academy report presents a comprehensive array of research about science communication, how people make decisions, and how these findings can promote more effective interaction between scientists and the public. The diversity of public opinion on some issues may be frustrating to many scientists, and many may think that if they just explain research and evidence more clearly and carefully, people will accept that GM foods are safe and human-induced climate change is really happening. If people made decisions about such issues rationally, this would be true. But many do not. Science communication researchers are finding that people mostly make these decisions for deep-seated social and cultural reasons. As researcher Dan Kahn sums it up: "People endorse whichever position reinforces their connection to others with whom they share important commitments. As a result, public debate about science is strikingly polarized. The same groups who disagree on 'cultural issues'—abortion, same-sex marriage and school prayer—also disagree on whether climate change is real" (Kahan 2010).

Professional Development for College Faculty

In the early '90s we learned that there were college faculty, particularly at smaller institutions, who also wanted to learn more about biotechnology, so we ran three general biotechnology workshops for them beginning in 1992. These were at a higher level than the secondary school teacher workshops, were two weeks long, and incorporated more sophisticated laboratory work. One workshop was taught by NC State faculty; and Helen Kreuzer and I each taught workshops in subsequent years in collaboration with other faculty. Later short workshops and one-day conferences for college faculty focused on:

- The bioprocess manufacturing industry;
- Developing educational projects with industry partners and integrating the teaching of professional career skills into curricula;
- Academic programs that blend science and business education, such as the Professional Science Master's degree.

Workforce Development

In addition to educating the public about biotechnology, and getting young people interested in careers in biotechnology, another main goal of the Education and Training Program was to support development of educational assets necessary to prepare the workforce needed by companies in North Carolina.

Between 1986 and 1996, growth in the U.S. biotechnology industry grew from 850 companies with 40,000 employees to 1,308 companies with 108,000 employees; and worldwide sales of products increased nearly 8.5-fold (Lee and

Burrill 1995). The number of new biopharmaceuticals approved by the FDA increased over ten-fold from four to 43 during this same decade (Annual Report: Top Biotech Companies 1996). It was clear that more and more new biopharmaceuticals were going to emerge from the lab and move into large-scale production. There were already 16 biomanufacturing sites in North Carolina, eight of which produced biopharmaceuticals. Adrianne recognized that bioprocessing could help replace the many manufacturing jobs being lost in North Carolina's traditional tobacco, textile, and furniture industries in the 1990's. It was high time for us to find out what the biomanufacturers' employment needs were, and what kinds of education and training were required.

Needs Assessment

I toured all the bioprocessing plants in the state and with my first visit became fascinated with the technology. It was amazing to see cultures of bacteria growing in three-story high tanks instead of in small flasks in a laboratory shaker bath; not to mention the maze of piping, pumps, valves and other equipment required to provide life support for the cells in the tanks and purify the desired product from the culture. I was hooked on bioprocessing.

I hired two contractors (Jean Chappell and Carol Schafer) and we gathered survey data and interviewed plant managers, gathering quantitative and qualitative information about each company's employment profile and hiring needs. We found that the most pressing personnel need of almost all these companies was for good process technicians—the people who operate and monitor the manufacturing equipment, and who constituted the largest fraction of the workforce. (These employees are sometimes called manufacturing technicians or process operators.) I spent three days shadowing technicians at the Ajinomoto amino acid production plant in Raleigh (which besides being quite informative was really fun). I summarized what we learned in the first *Window on the Workplace* publication (Kennedy 1997).

The Challenge

These jobs paid well, but it was hard to find people who had some relevant manufacturing or other experience, as well as appropriate employability attributes and at least a high school education. Willingness to do shift work also was a requirement, since most of the companies operated around the clock. Companies working to the Good Manufacturing Practice guidelines required by the FDA for pharmaceutical production had a particular need for employees with prior experience in this area. In the absence of a good candidate pool at that time, companies often were hiring away each other's technicians. Adrianne thought the answer might be in providing specific training for the many workers getting laid off from textile plants, but discovered that the vast majority did not have a high school education. There were a few related Associate of Applied Science (AAS) degree programs at the community colleges, but the annual output of graduates was insufficient to meet the need.

The Solution. Ms. Joanne Steiner, human resource manager at Novozymes North America (which produces industrial enzymes from microbial cultures), had developed a creative approach to their hiring problem. She arranged with the local Vance-Granville Community College to host a short course for people who were interested in process technician jobs. The class was taught by company employees and provided an introduction to the company, its manufacturing processes, and some basic science and math. People who completed the course satisfactorily were invited to interview for jobs.

This became the model for our solution to the process technician problem. What if we could develop a course that provided basic training suitable for a broad assortment of companies and could be offered anywhere in the state? We did exactly that in collaboration with industry and the community colleges. It would be called *BioWork* (2011).

Developing BioWork

The course objective was to teach people interested in jobs in the industry what to expect in the manufacturing environment (and whether this would be a job they really wanted), give them enough background knowledge to learn specific tasks more quickly from their first day on the job, and provide them with a basic understanding of what was going on in the processes they would run. I developed a draft competency list that went through two cycles of review by industry employees. A compendium of competencies for chemical process technicians produced by the American Chemical Society was an invaluable reference (Hofstader and Chapman 1997).

The final content of *BioWork* included:

- Descriptions of job environments and expectations of employees; and workplace safety;
- Relevant basics of chemistry, physics, microbiology, and math;
- Common laboratory procedures used in testing product samples during processing;
- FDA-required Good Manufacturing Practice guidelines and record-keeping;
- Effective communication and teamwork;
- Typical chemical, pharmaceutical and biopharmaceutical production processes and equipment; plant utilities and waste treatment;
- Automated process control systems and instrumentation for process monitoring.

I hired a team of two capable technical writers, Sandy Thomas and Pat Hill, and artist Joe Delmonico to produce the text, and Maria and I also wrote parts of it. Reading level was set at 10th grade. The designed total class time is 116–120 h of lecture and lab, delivered in one semester. We had course units reviewed by industry experts and then field-tested the course with both new and experienced employees at Novozymes North America, Inc. and Biogen. We got more helpful feedback, made revisions; and started publication in 2001.

By the end of 2012, the North Carolina Community College System had recorded over 7,000 enrollments, according to data provided to the NCBiotech Library. (Enrollment does not indicate course completion.) This number likely includes both public classes at the colleges and customized industrial training conducted by the colleges for individual companies. Companies started to make *BioWork* a hiring requirement, or used it as part of their training of new hires. *BioWork* is still being taught wherever needed across the state.

Comprehensive Workforce Development: NCBioImpact

BioWork was an effective solution for one part of the workforce demand, but it was just the first of a more comprehensive set of assets for workforce training that North Carolina would build. Companies wanted a better pipeline of qualified applicants for many other kinds of jobs, as well as more hands-on training with appropriate equipment for process technicians.

The Biomanufacturing Forum is a group of biomanufacturing plant site managers that meets regularly to discuss common interests and needs. It is part of the state's industry trade group, NCBIO (www.ncbioscience.net). Members of the Forum worked with educators to develop a plan for building facilities and degree programs that would address their personnel needs in diverse areas from process development to manufacturing. Our second study of the industry, *Window on the Workplace 2003* (Kennedy 2003) documented the then-current trends in biomanufacturing sector growth, hiring, and employee education requirements; and provided the data to support a request by NCBIO to the state's Golden LEAF foundation for funding (http://www.goldenleaf.org).

Golden LEAF provided over $63.3 million. Industry provided over $13 million in-kind, in the forms of donated equipment and company engineers' time to assist with the design and construction of new facilities. A consortium of institutions including two universities, the Community College System, NCBIO and the Biotechnology Center was formed to implement and guide the new initiative. This consortium was called NCBioImpact (http://www.ncbioimpact.org).

A curriculum committee of faculty and industry representatives was convened and had the idea of using job descriptions for the six most common entry-level positions in the industry as a framework for identifying the required knowledge and skill base for new and existing degree programs. (We defined an entry-level position as one suitable for persons with appropriate education but no prior work experience in the industry.) John Balchunas, working with focus groups of employees for each job, organized a list of detailed knowledge and skill standards to guide curriculum development in the colleges and universities. This compendium was *The Model Employee: Preparation for Careers in the Biopharmaceutical Industry,* and was distributed to college faculty around the state (Balchunas 2005). This text includes typical hiring requirements and job descriptions as well as career development pathways for each position. The aforementioned *Career Pathways* booklet for secondary school teachers built from this foundation.

The components of NCBioImpact are:

- At North Carolina State University in Raleigh, a new building, the Biomanufacturing Training and Education Center (BTEC) housing a pilot-scale bioprocessing plant and cleanrooms for hands-on training in both university and community college level programs;
- At North Carolina Central University, a historically minority institution in Durham, a new building housing research and teaching laboratories for new BS and MS degree programs in Pharmaceutical Science. This initiative is called the Biomanufacturing Research Institute and Technology Enterprise (BRITE);
- In the North Carolina Community College System, a new organizational unit, NCBioNetwork, to provide courses and training services at campuses and companies across the state; a mobile teaching laboratory; and classrooms and laboratories at the BTEC.

A Win–Win Solution

Joy Callahan, Dean of Economic and Workforce Development at Johnston Community College in Clayton, NC, told me in an interview (29 Feb 2016) about the history of another industry collaboration with government and educators in 2005. Two major biologics manufacturers, Novo Nordisk and Grifols, worked with county officials to build a training facility in their area. Novo Nordisk donated land. The county obtained funding from Golden LEAF and the U.S. Department of Agriculture for construction. County taxes paid by companies located in a designated area cover the operating expenses of the training center. Johnston County Community College operates the center, providing *BioWork* and specialized short courses for company employees and the general public—some of whom may become employees. The center also provides coursework for an AAS degree in Bioprocessing Technology. In 2016, both companies announced new expansions. Novo Nordisk's new $1.8B facility will require 700 new employees—a doubling of its workforce. Grifols recently announced a new $210 M plant expansion less than a year after constructing another new building. See http://www.johnstoncc.edu/about/workforce; http://www. ncbiotech.org/article/novo-nordisk-breaks-ground-18b-drug-plant-clayton/ 168951); http://www.ncbiotech.org/article/grifols-adding-210m-expansion-clayton-campus/166346.

Education Outcomes for Industry

John's subsequent studies of biomanufacturing employment indicated significant changes in the educational profile of the workforce. He identified a group of eight companies with 4,686 employees for which we had survey data from both 2002 and 2007 (indicated as 2008 in *Window on the Workplace 2012*). Most of these companies are biopharmaceutical manufacturers. During this period the number of workers in the surveyed population with *only* a high school education had decreased by 27%, and the number of workers with a high school diploma *plus* some type of certification had increased by 237%; a remarkable increase that he attributed to the availability of *BioWork* as well as a variety of other continuing education courses offered by the community colleges (particularly at the BTEC). The number of employees with master's degrees had increased by 157%. Employees with bachelor's or associate degrees made up just over half the workforce (53–56%) in both survey years (Balchunas and Kennedy 2012a).

Even in the aftermath of the 2008–09 recession, growth of biomanufacturing in North Carolina continued. The availability of qualified employees was a plus, enriched not only by NCBioImpact programs but by in-migration of qualified people attracted by industry growth. In 2007 biomanufacturers were hiring 30% of their employees from out of state. By 2012, they were hiring only 10% of their employees from out of state—usually people with highly specialized expertise (Balchunas and Kennedy 2012b).

New Hiring Patterns and Implications for Education

The shift to leaner operations and more highly educated employees was beginning to be a general trend in business even before the recession (North Carolina Commission on Workforce Development 2011a). Speakers at a 2016 forum at North Carolina State University projected future strengthening of this pattern globally due to increasing automation and computerization—a trend that could displace not only lower skilled workers but also some knowledge workers (See www.iei.ncsu.edu/futurework/resources-materials). In 2011, a North Carolina Workforce Commission study projected that 42% of new jobs created in the state over the next decade would require at least some post-secondary education; and noted that "A high school diploma alone will no longer offer even a remote pathway for future success." Many jobs, often in manufacturing, that used to provide a middle-class wage have disappeared; and emerging new jobs at the same wage level, for example in health care, require specialized training (North Carolina Commission on Workforce Development 2011b).

Secondary school teachers likely are aware of these trends, especially in rural regions where jobs have been lost and economic development infrastructure to attract new jobs is often limited. It is critical for teachers to let students know that they will need to pursue further education or job training. Parents need to know this, too, but in high poverty areas (urban or rural) it can be impossible for many of them to afford. Fortunately, North Carolina's community college system comprising 58 colleges, many with more than one location, serve every county in the state and are a low-cost solution for many families.

Teachers have been responding to the changing demands of the workplace by developing student projects that often are relevant to their community and its environment. These and other kinds of project-based learning at their best can inculcate essential workplace skills such as creativity, problem-solving, entrepreneurship, effective communication, collaboration and networking. The Center has provided grant funding to support some of these efforts.

Education Enhancement Grants

Beginning in 1991 the Education and Training Program provided grants for up to $100,000 to schools, colleges, universities, and non-profit organizations to support biotechnology education projects. Initiating this program was one of my first jobs when I joined the Center. Later, Bill took over management.

Awards were made for teaching laboratory equipment, new course and curriculum development, faculty professional development, and other kinds of projects. Proposals were reviewed by ETP staff and experts both in and out-of-state, and funding recommendations were made by panels of 10–12 North Carolina science teachers, college and university faculty, and industrial scientists.

In later years, we also had two smaller funding programs:

- Mini-Grants provided awards of up to $6,000 to school teachers for equipment, supplies, or course development.
- Undergraduate Research Fellowships provided $5,000 awards to support students working on biotechnology projects. Cotton Incorporated donated funding for two fellowships per year in agriculture-related research.

As of 2011, we had distributed a total of over $7.4 M, including 77 awards to secondary schools or school systems. Examples of education grant awards made during the last few years of the program are listed below. A complete list of all Center grant awards made for support of science and education from 2011 to 2015 can be found at: http://www.ncbiotech.org/past-awards.

Four-Year Colleges and Universities

North Carolina State University: Collaborative project among faculty at Davidson College and Lenoir-Rhyne University with North Carolina State University scientists at the North Carolina Research Campus in Kannapolis to develop new plant genomics courses for undergraduates; and develop tutorials and web interfaces for students to learn to annotate blueberry genomic sequences. www. northcarolina.edu/nc-research-campus-kannapolis; www.plantsforhumanhealth. ncsu.edu.

UNC-Charlotte: Acquisition of next generation sequencing equipment to support a graduate minor program in Genomics and Bioinformatics. Graduates from this program acquire expertise in laboratory methods as well as bioinformatics.

UNC-Wilmington: Integration of molecular techniques across the biology curriculum in labs for several courses. Labs in five courses will benefit from a real-time PCR detection system, imaging software for a confocal microscope, thermal cyclers, a gel documentation system, and a spectrophotometer. This is a nice example of projects we funded at many institutions for modernizing course content and upgrading teaching lab equipment.

Secondary Schools

Avery County Public Schools: Initiation of a Biotechnology and Agriscience Research course, with major emphasis on plant tissue culture. Avery County, billed as the "Christmas tree capital of the world," is in a high biodiversity area. Nurserymen who profit from selling indigenous horticultural varieties wish to protect the area from overharvesting, and need to utilize micropropagation techniques. The high school course introduces students to this technology and further training is available at the local community college.

Polk County High School: Development of student research opportunity. "The Magnolia Detectives Project" was one of the review panel's favorite proposals. Teacher Jennifer Allsbrook noticed an unusual stand of sweetbay magnolia growing in her county in western North Carolina. It usually grows only in coastal areas. She thought it would be interesting for her class to perform genetic analyses to determine the closest relatives of these trees. We initially gave her a professional development award to travel for consultation with experts and the following year we funded classroom implementation.

Workforce Development

Appalachian State University: Development of two courses within a new Fermentation Science degree program. This is the first such program on the East Coast, and supports a growing concentration of craft brewing industry in the mountains as well as other parts of the state. The program earns supplemental income from its Ivory Tower Brewery sales (Balchunas and Kennedy 2012c).

East Carolina University: Development of two-day course modules in separation technology and aseptic manufacturing for engineering majors. Modules are taught at the BTEC at NC State University to give students hands-on experience in a manufacturing-like environment. The bioprocessing concentration is an asset for the growing pharmaceutical cluster in eastern NC. An earlier award to the ECU Chemistry Department supported a quality control training laboratory set up and operated according to industry Good Laboratory Practice standards. Local pharmaceutical industry scientists teach the classes.

North Carolina State University: Creation of a new nanotechnology manufacturing course at the BTEC. This course covers the manufacture, characterization, and applications of nanomaterials and provides hands-on experience with their preparation—an interesting complement to the BTEC's primary emphasis on bioprocess manufacturing.

Non-profit Organization

North Carolina Museum of Natural Sciences: Acquisition of microscopes suitable for use by young people and associated projection equipment for the Genomics and Microbiology "Investigate Lab." This is one of three demonstration laboratories in the Nature Research Center, a new wing of the museum that opened in 2012. Lab programs were projected to reach 14,000 visitors during the first year of operation. http://www.naturalsciences.org.

Afterword

Norris Tolson, President and CEO of the Center from 2007 to 2014, summed up his view of the work of the Education and Training Program as follows:

> The Center had and still has an essential education job to accomplish for a wide range of stakeholders. I believe industry understood best how important the Center's education programs were to their success. Evidence for this is the continued funding of ag biotech workshops by a group of companies after the Center's education unit had to be closed. Worker training is, of course, a critical need; and is one of the strongest recruiting tools that North Carolina has for biotech business. But so is general public understanding of what biotech offers the world. I cite as a 'gold standard' the program to equip teachers to be a vital link to students, who in turn can be a conduit of information to their parents and friends—all of whom are consumers. The contributions of effective education to North Carolina's status as a global biotech hub have been *crucial* and will continue to be so.

In closing I would like to note again that our work in the Education and Training Program took place within the context of the Center's activities as a whole. Consistent strategic investment in capacity building and relationship building over more than 30 years has paid off handsomely. The combination of support for education, academic research, small company startup and large company recruitment has had a synergistic impact on the growth of the industry in North Carolina greater than what any of those efforts alone might have accomplished. It was exciting and rewarding to have been part of this enterprise.

References

Annual Report: Top Biotech Companies (1996) MedAdNews 15:32

Balchunas J (2005) The model employee: preparation for careers in the biopharmaceutical industry. North Carolina Biotechnology Center, Durham, NC. http://www.ncbiotech.org/workforce-education/k12-education-support. Accessed 6 July 2016

Balchunas J, Kennedy K (2012a) Window on the workplace 2012: North Carolina's biomanufacturing and pharmaceutical-manufacturing workforce. p 34. North Carolina Biotechnology Center, Durham, NC. http://www.ncbiotech.org/workforce-education/k12-education-support. Accessed 6 Jul 2016

Balchunas J, Kennedy K (2012b) Window on the workplace 2012: North Carolina's biomanufacturing and pharmaceutical-manufacturing workforce, p 38

Balchunas J, Kennedy K (2012c) Window on the workplace 2012: North Carolina's biomanufacturing and pharmaceutical-manufacturing workforce, p 16.

Balchunas J, Omohundro J (2006) Career pathways: focus on biotechnology. North Carolina Department of Public Instruction, Raleigh NC and North Carolina Biotechnology Center, Durham, NC. http://www.ncbiotech.org/workforce-education/k12-education-support and http://www.ncbiotech.org/workforce-education/biotech-career-guide. Accessed 10 April 2016

Battelle Technology Partnership Practice (2014a) 2014 Evidence and opportunity: impact of life sciences in North Carolina, p xviii. http://www.ncbiotech.org/sites/default/files/pages/2014%20Battelle%20Report_Full_0.pdf

Battelle Technology Partnership Practice (2014b) 2014 Evidence and opportunity: impact of life sciences in North Carolina, p x

Battelle Technology Partnership Practice (2014c) 2014 Evidence and opportunity: impact of life sciences in North Carolina, pp 3–6

Battelle Technology Partnership Practice (2014d) 2014 Evidence and opportunity: impact of life sciences in North Carolina, pp 1–2

BioWork: An Introductory Course for Process Technicians (2nd edition 2011) North Carolina Biotechnology Center, Durham, NC

Ernst and Young (2004) Resurgence: the Americas perspective–global biotechnology report

Hofstader R, Chapman K (1997) Foundations for excellence in the chemical process industries: voluntary industry standards for chemical process industries technical workers. American Chemical Society, Washington DC

Kahan D (2010) Fixing the communications failure. Nature 463:296–297

Kennedy K (1997) Window on the workplace: workforce training needs for North Carolina's bioprocessing industry. North Carolina Biotechnology Center, Durham NC

Kennedy K (2003) Window on the workplace 2003: a training needs assessment for the biomanufacturing workforce. North Carolina Biotechnology Center, Durham, NC

Kreuzer H, Massey A (2005) Biology and biotechnology: science, applications, and issues. ASM Press, Washington DC

Kreuzer Helen, Massey Adrianne (2008) Molecular biology and biotechnology: a guide for teachers, 3rd edn. ASM Press, Washington DC

Lee K, Burrill S (1995) Biotech '96: pursuing sustainability: the tenth industry annual report. Ernst & Young, LLP

National Academy of Sciences (2015) NAS public engagement on genetically modified organisms: when science and citizens connect: a workshop summary. National Academies Press, Washington DC

North Carolina Commission ̄on Workforce Development (2011a) State of the North Carolina workforce 2011–2020. pp iii–iv. https://www.nccommerce.com/Portals/11/Documents/Reports/2011%20SOTW%20Full%20Final%20Report%2052311%20909am.pdf. Accessed 07 Jul 2016

North Carolina Commission on Workforce Development (2011b) State of the North Carolina workforce 2011–2020, pp 21, 56

Ronald P, Adamchak R (2010) Tomorrow's table: organic farming, genetics, and the future of food. Oxford University Press

Shank K, Niebauer A (2016) North Carolina Ag biotech: economic growth report. https://www.ncbiotech.org/sites/default/files/pages/AgBiotechEconomicGrowthReport_April%202016%20FINAL.pdf. Accessed 29 June 2016

Africa's Fight for Freedom to Innovate and the Early Signs of Embracing Biotechnology Especially Genetically Modified (GM) Foods

Florence M. Wambugu

Background

On 1 July 1999, my article *Why Africa needs agricultural biotech* was published by Nature.[1] In retrospect, this was a watershed article that was to later open a vicious battle for the hearts and minds of the African people regarding the genetic modified (GM) technology. My thesis was simple: Africa did not just need the GM technology, it needed it urgently. I argued and urged Africa to enthusiastically join the biotechnology revolution. "There is urgent need for the development and use of agricultural biotechnology in Africa to help to counter famine, environmental degradation and poverty," I argued.

By this time, I was the Director of the Africa Regional Office of the International Service for the Acquisition of Agri-Biotech Applications (ISAAA AfriCentre), which I had established in Nairobi. The *Nature* article provided a watershed moment; it opened the floodgates for Africa to seriously evaluate where the continent stood on the GM debate. The attacks by the anti-GM lobby especially from Europe came fast and furious.

Clutching straws, those opposed to the technology accused me of claiming that the GM technology was the only answer to world hunger. In the cacophony of noise that followed, anti-GM activists did not want to hear that GM crops—within a bouquet of other agricultural technologies—had the potential to contribute to the reduction of poverty and hunger in the developing countries. The *Nature* article thrust me into the centre of the GM technology debate in Africa. This came at a time when people outside the continent were speaking for Africa. Now the Africans had a voice.

---

[1] http://www.nature.com/nature/journal/v400/n6739/full/400015a0.html.

F.M. Wambugu (✉)
Africa Harvest Biotech Foundation International (AHBFI), Nairobi, Kenya
e-mail: fwambugu@africaharvest.org

© Springer International Publishing AG 2017
L.S. Privalle (ed.), *Women in Sustainable Agriculture and Food Biotechnology*,
Women in Engineering and Science, DOI 10.1007/978-3-319-52201-2_8

About 2 years after the *Nature* article was published, the then Kenya Agricultural Research Institute (KARI), approved trials to develop a virus resistant sweet potato. Within 3 years, anti-GM activists were trumpeting the trials as a failure. Those who know that it takes over 10 years to develop and commercialize a GM crop were petrified; but the anti-GM activists did not want the facts or technical details to get in their way.

The anti-GM activists quoted two researchers, Dr Francis Nang'ayo, and Dr Ben Odhiambo saying: "There is no demonstrated advantage arising from genetic transformation using the initial gene construct." The fact that initial GM work was done at the Monsanto labs in the US made an excellent narrative. Monsanto had developed a coat protein responsible for virus resistance and donated it to KARI, royalty free, to use in its sweet potato improvement programme.

"The transgenic material did not quite withstand virus challenge in the field," a report by Drs Nang'ayo, and Odhiambo said, casting doubt on whether the gene expression was adequate or it failed to address the diversity of virus in Kenya. The "failed" experiment "corresponded" with an earlier study released by the Third World Network Africa, a well-known anti-GM organization. Their study, "Genetically Modified Crops and Sustainable Poverty Alleviation in Sub-Saharan Africa: An Assessment of Current Evidence," by Aaron deGrassi, of the Institute of Development Studies, University of Sussex, UK, warned that the GM sweet potato introduced in Kenya did not address the crop's major problem—weevils. The TWN study offered "new evidence" against claims of the miracle potential of GM crops for dealing with famine and poverty in Africa.

At the time the stories about the "failure" of the GM sweet potato were unravelling, I had moved from ISAAA and started Africa Harvest, a non-profit organization whose vision is to be a lead contributor to fighting poverty, hunger and malnutrition. The GM sweet potato project started as my Ph.D. project at the University of Bath in the United Kingdom. Later, on a post-doctoral assignment in Monsanto through a joint USAID Sponsorship, I did further work on the GM sweet potato. The reports of failure were therefore of concern to me, not just because my name was being dragged in the mud, but because of the falsehoods peddled by the anti-GM groups.

I came out strongly to demonstrate that the reports had been completely mis-interpreted and distorted. Contrary to what anti-GM activists were saying, the GM sweet potato has been a success in many ways. As indicated, it takes over a decade to initiate, develop and commercialize a GM crop; under careful biosafety approval processes, the molecular biology research laboratory work moves to confined greenhouse trials (CGT) and, subject to further approvals, to Confined Field trials (CFTs) before Open Field Trials (OFTs), which is the first part of the deregulation process.

Typically, the first generation products are not intended for commercialization. The GM sweet potato variety tested in Kenya was meant to develop a genetic transformation system which did not exist globally. It contained a reporter Gus gene which is a 'tell-tale' gene commonly used to indicate to scientists whether a plant is indeed transformed. Reporter genes like the Gus gene are not included in final commercial products and are out-bred once the final product has been established. The sweet potato variety produced in Kenya's was the first generation product developed for the system. The field trials results were meant to identify the level of protection needed for the final product in Kenya. The purpose of the field trials was also to shed some light on how to improve the system used to transform the sweet potato. The "failure" was merely a *scientific finding* that indicated the extent to which sweet potatoes were vulnerable to disease in the region in which the trials were carried out.

As the anti-GM activists were obfuscating the issues, scientists, in anticipation of field trials' results, were already working a second generation product which included a gene construct from the most virulent Kenyan potato virus strain. The Muguga virus strain had been identified after extensive screening. Future research was designed to produce a second generation GM sweet potato variety equipped with double protection. The protective feature of this GM variety would have both the coat protein (CP) gene and its replicase gene which had the special ability to prevent the sweet potato feathery mottle virus (SPFMV) from replicating upon infection, thereby rendering the virus harmless. An additional cloning site to the gene construct had been made, which would make it much easier for scientists to add the gene that gives it resistance to weevils, if and when this was discovered. At the time, it was thought that the final GM sweet potato product would be tailor made for African environmental conditions.

Although it was never commercialized, in many ways, the GM sweet potato project more than achieved its goals, including the development of a scientifically sound genetic transformation system for sweet potato. Being the first GM crop variety in Sub-Saharan Africa, the pioneering nature of the project demanded adherence to strict international standards. The trials were carried out after close consultation and in close collaboration with the rural communities where the sites were located.

In less than 15 years since the project was started, many Kenyan scientists have been trained through the project and many other GM projects have been initiated and are ongoing. The human and infrastructure capacity development created the starting point that has over the years, enabled the country to define its nature of engagement with the GM technology. Kenya now has a bio-transformation lab where other crops—other than the sweet potato—have been developed. The lab puts Kenya in a position to form vital collaborations and further build the country's scientific, and more specifically, GM technology capacity.

The country is also a beacon of light in the region with regard to biosafety and GM technologies. Organizations such as Kenya Plant Health Inspection services (KEPHIS) developed initial expertise and experience on how to regulate GM Crop field trials, out of the GM sweet potato project. KEPHIS monitored all field trials, collected and analysed data to ensure compliance with internationally accepted standards. Today, the country has a fully-fledged, pro-science biosafety law and an operational National Biosafety Authority (NBA). At the time of writing, NBA is evaluating several applications on GM Crops that could see the commercialization of GM crops.

Leveraging South Africa's Early Foray into the GM Technology for the Rest of the Continent

Africa's history with the GM technology is less than two decades. Although research was on-going in various countries, it was not until the passage of the Biosafety law in parliament allowing commercialization in South Africa (GMO Act, 1997) that the continent can claim to have staked its claim. Shortly after passage of the law, South Africa commercialized Line 531/Bollgard (Bt. Cotton) and MON810/Yieldgard maize. The anti-GM activists may have been caught unaware by the momentum building in South Africa.

My *Nature* article, coming 2 years after passage of the South African law and commercialization of both cotton and maize seemed to have provoked the hornets' nest. Its premise was that this technology was needed, not just in South Africa, but throughout the continent. In South Africa, GM activism increased and in 2003, The Biowatch Trust appealed a decision against the Executive Council to authorise commercial growing of maize event Bt11. They lost the case. The upholding of the decision of the Executive Council confirmed the effectiveness of the country's biosafety regulatory system.[2]

The following year, Biowatch sued the National Department of Agriculture (NDA)—under which the GMO Council falls—demanding access to all documentation relating to the administration of the GMO Act. The NDA was open to providing some of the information while protecting the confidential business information contained in many of the documentation requested at that point. Biowatch considered this information as inadequate and therefore sought relief from the High Court. Although the court ordered the directorate to provide the information requested, it is important to note that the ruling confirmed the department's responsibility to protect confidential information.

These interactions framed the early GM debate in Africa. It was about this time that I left ISAAA to start Africa Harvest Biotech Foundation International (AHBFI, or simply Africa Harvest). In 2002, barely months after the Foundation was launched, we were involved in the United Nations (UN) World Summit on Sustainable Development (WSSD). Prior to the meeting, I made a presentation on African Biotechnology issues for WSSD Regional Parliamentarians meeting held at the White Sands Hotel in Mombasa, Kenya.

At the Johannesburg meeting, I "stressed the need to combine the use of biotechnology with good governance in Africa."[3] I also proposed that Africa look to regional markets as an alternative to Europe, which was non-committal or opposed to the GM technology. More significantly, I presented an Africa biotech stakeholders position on the GM technology, declaring it immoral for African

---

[2]Annual Report of the Executive Council of the Genetically Modified Organisms Act, 1997 (Act No. 15 of 1997) for the period 2004/05 http://www.nda.agric.za/docs/GeneticResources/gmo%20res%20act%20.pdf.

[3]http://www.iisd.ca/2002/wssd/enbots/pdf/enbots1006e.pdf.

governments to reject genetically modified (GM) crops and foods when people were dying from hunger. This was in reference to Zambia's rejection of food aid containing GMOs.[4] We later travelled to Zambia and held high-level meetings regarding this issue and were assured that it has been blown out of proportion, but over time, because of politics, the government hardened its position. In recent years, Zambia has build capacity and could be an important player in the technology in future.

Given the context within which Africa Harvest was established, our focus was to ensure a coherent African position on the GM technology. We adopted a three-pronged approach: international, pan-African outreach and focus on specific countries, where we saw potential for fast-tracking the acceptance of the technology. At the international level, we, for example, submitted a statement to the Committee on Agriculture United States House of Representatives Hearing[5] on "Review of Artificial Barriers to United States Agricultural Trade and Foreign Food Assistance."

At the pan-African level, we focused on ensuring a progressive position on the GM technology at the highest levels. As a founder of Executive Committee member of Forum for Agricultural Research in Africa (FARA), we were involved in the drafting of the Comprehensive African Agricultural Development Program (CAADP) Pillar 4 on agricultural research. Within it, biotech and the GM technology were identified among the key drivers of African agricultural research. This was a precursor of the African Union (AU) and the New Partnership for Africa's Development (NEPAD) position on the GM technology captured in the *Report of the High-Level African Panel on Modern Biotechnology.*[6]

On a country-to-country level, we worked, with support from Croplife International (CLI), in African countries that we thought had the greatest potential to adopt biosafety regulations, paving the way for acceptance and commercialization of the technology. Working in an extremely challenging and complex environment, our efforts have focused on ensuring policy makers create an enabling policy framework for the GM technology. We work with different stakeholders to avoid delays in the adoption of the technology on the continent.

Among Africa Harvest successes—working with a myriad of international and local stakeholders—is the passage of pro-science Biosafety laws in Kenya, Nigeria and Ghana. Although only four countries (South Africa, Burkina Faso, Sudan and Egypt) have commercialized GM crops, many others have the necessary legal framework and human capacity to do so. More specifically, Ghana and Kenya are evaluating applications for various crops that could lead to commercialization. I have argued that the key to fast-tracking the GM technology on the continent is political will, because the GM technology regulatory system has become highly

---

[4]http://www.telegraph.co.uk/news/worldnews/africaandindianocean/zambia/1411713/Starving-Zambia-rejects-Americas-GM-maize.html.

[5]http://www3.bio.org/foodag/action/20030326.asp.

[6]http://belfercenter.ksg.harvard.edu/files/freedom_innovate_au-nepad_aug2007.pdf.

political. Purely establishing the food and environmental safety of a GM-Product in Africa does not lead to regulatory approval as the EU position has to be considered as well as other external factors. Therefore it is obvious to conclude that "Unless and until as African country has the political will and support for biotechnology application, investments made in biotechnology will not be fully realised.[7]"

Using Tissue Culture (TC) Biotechnology to Show Case the Benefits of Biotechnology to Smallholder Farmers in Africa. The Case of Tissue Culture Banana

Securing the political will of African leaders and policy makers requires a deep understanding of their concerns and fears. Since most politicians have to resolve complex, poverty-related needs, it's imperative to step-down the scientific language and to demonstrate the benefits of the biotechnology. This has been an important part of our strategy.

This challenge has also allowed me to continue focusing on my passion, which is to help increase agricultural productivity for resource-poor farmers. For me, GM technology is only one of the tools in a large arsenal of technologies available to scientists and farmers. Of course conventional technologies still have an important role to play; What Africa needs is the freedom to chose whatever agricultural technologies will address its challenges.

We have also found it critical to push for agricultural R&D to incorporate home-grown ideas and innovations. Forced by years of limited success, development players are now searching for how best to tap farmers' indigenous knowledge and innovations. Africa Harvest's Tissue Culture Banana projects have captured the imagination of farmers, politicians and policy makers and become a smooth entry point from a discussion on biotechnology to the GM technology.

In a country like Kenya, where approximately 75% of the total population (33 million of the estimated 45 million) lives in the rural areas[8] and the most important economic activity is smallholder farming, discussions about the GM technology can be removed from real or immediate needs. The challenge is compounded by the fact that in most African countries, most of the population lives in abject poverty with incomes of less than US$1 per day.

Africa Harvest vision and mission strategy is focused on addressing the issue of poverty, hunger and low incomes[9] using existing convention technologies, before

---

[7]Biotechnology in Africa: Emergence, Initiatives and Future, Editors: Wambugu, Florence and Kamanga, Daniel (Eds.).

[8]Kenya's Central Bureau of Statistics (CBS), 2004.

[9]Low incomes translate to poverty and hunger, compounded by a vision cycle of poor crop yields, poorly functioning markets and inefficient pricing. Eventually, these vulnerable groups are unable to purchase food and have to receive food assistance during times of severe food shortages.

or alongside the discussion on the GM technology. The problem of low incomes is linked to value-chain inefficiencies[10] especially with regard to food and 'orphan crops' such as banana and sorghum. Addressing the issues facing the smallholder farmers has given Africa Harvest the required trust to discuss opportunities offered by the GM technology.

For example, the fact that banana is the most widely grown fruit in Eastern Africa means[11] significant investments have been made in the tissue culture (TC) banana technology transfer. Africa Harvest is a trusted technology dissemi-nator. Over the last 10 years, the International Development Research Centre (IDRC), the Rockefeller Foundation and DuPont Pioneer have jointly made investments through funding Africa Harvest to the banana sub-sector. Generally, funding has focused on technology transfer to increasing banana production accompanying interventions along the whole value chain. In 2002, the Rockefeller Foundation expanded its support to the banana sub-sector to include access to markets and 2 years later, DuPont Pioneer funded an Africa Harvest project, which focused on the Whole Value Chain (WVC) strategy.

The TC banana project clearly demonstrates that technology seedlings derived from the lab can benefit small holder farmers positively synthesizing key stake-holders in agriculture—local and national leaders, politicians and policy makers—appreciate the important role biotechnology can play if well targeted and is need based. Partnership with multinational companies like DuPont Pioneer for the common good with a non GM-crop also demonstrated good will against the accusations of "always seeking opportunity to sneak in GM Crops".

The ABS Project's Role in Consolidating Africa's Experiences and Lessons Learnt in GM Technology

The Africa Biofortified Sorghum (ABS) Project (www.biosorghum.org)—with the vision to develop sorghum varieties targeting critical nutritional deficiencies—has probably made the biggest difference in biotech experiences & lessons learnt in Africa. The project was initially funded by the Bill and Melinda Gates Foundation (BMGF) from 2005 to 2012. During the period, the project included a consortium of nine partnership institutions in Africa and USA, many collaborator organizations & stakeholders. The ABS included a network of over 70 scientists spread over five countries in Africa (South Africa, Kenya, Burkina Faso, Egypt & Nigeria), and

---

[10]Limited access to, and when available, high cost of credit as well as lack of information and poorly functioning markets (International Monetary Fund, IMF, 2005).

[11]Total area under banana: 40,000 ha (MOA, 2004).

USA as well as in several regional bodies. Other funding has been received from the Howard G. Buffet Foundation through the Danforth Center, with additional in-kind and monetary support from DuPont Pioneer.

The project has made major technological breakthroughs that include achieving 70 µg/g of β-carotene accumulation, 140 times higher than that of wild-type sorghum (TX430) and more than double than that of Golden Rice-2®. The project has also demonstrated that co-expression of vitamin E could improve stability of β-carotene in sorghum during storage and increase the β-carotene half-life in grain from about four to 10 weeks. Based on conservative estimates, currently achievable biofortified sorghum has the potential to contribute from 35 to 60% of the recommended daily allowance (RDA) of vitamin A for children in Africa. The focus has now expanded to achieving enhanced bioavailability of zinc and iron and improving protein digestibility.

This success should be viewed against the fact that sorghum is the world's fifth most important cereal and serves as the major food staple for the world's poorest 300 million people who reside largely in sub-Saharan Africa and southern India. Among other major achievements, the ABS project has:

- Increased pro-vitamin A content in sorghum
- Increased zinc and iron bioavailability through phytate reduction
- Reduced time required for sorghum transformation by 60% to 4 months
- Increased transformation success rate in sorghum 100 times over previous capabilities
- Improved protein digestibility levels after cooking
- Adhered to transgenic biosafety principles and best practices specific to Africa
- Conducted nearly a dozen confined field trials in Kenya and Nigeria with strict adherence to stewardship and compliance protocols
- Continued capacity building among African scientists and researchers.

As a public-private consortium actively working to improve the health of millions of people who rely on sorghum as their primary diet by enhancing its nutritional quality through biofortification, the project has helped demonstrate potential benefits of the GM technology. Working with DuPont Pioneer, Africa Harvest has helped build scientific and biosafety leadership in several African countries. The project has trained over 20 African Scientists in the USA on advanced GM technologies and other relevant skills including biosafety and regulatory assessment. Although the consortium has changed over the years, the core partners remain; these are the Kenya Agricultural and Livestock Research Organization (KALRO), Nigeria's National Biotechnology Development Agency (NABDA) and the Institute for Agricultural Research (IAR) and DuPont & Africa Harvest. It is important to acknowledge the early contribution of South Africa's Agricultural Research Council (ARC), the Council for Scientific and Industrial Research (CSIR) and the University of Pretoria. The University of California Berkerly (UCB) was also involved in the early stages of the project. In Burkina Faso, the Institut de l'Environnement et de Recherches Agricoles (INERA) was also involved in the project; the institute remains interested in rejoining the project, subject to strategic and funding imperatives.

The ABS Project, is a forerunner to several other African GM projects, and has helped to consolidate the enormous work of many scientists, local and international organizations. Many individuals have fought and continued to fight for GM technology acceptance in Africa. There is no doubt there is light at the end of the tunnel as several Africa Countries have commercialized GM crops and will continue to do in the future, gaining from the now 20 years of experience since the first GM crop was planted in 1996.

Sustainable Agriculture and Biotechnology—A Woman Grower's Perspective

Carrie Mess

I may be the most unlikely person to be writing a chapter in this book. I have never worked in a lab, I never even took chemistry in high school. In fact, all through school I struggled in math and science, while I could understand the concepts in my earth science and animal science classes, chemistry and algebra would cause my head to swim and for me to feel like I wasn't very smart. It wasn't until after I graduated high school that I found out that I had undiagnosed learning disabilities that kept me from learning the building blocks that build the concepts taught in those classes and beyond. I thought I didn't like science, I thought I was horrible at math. It turns out, I was wrong. Today my office is a dairy barn or tractor cab and although my job doesn't come with a white coat, the work done by the other women in this book, is also my work.

Looking at me today on the farm, it's easy to imagine that I've always been a farmer and that I was raised with the cycles of planting and harvesting, birth and death as part of my life. However, while I may look and act the part of a farmer, I was raised in the city and had no real connection to agriculture until after I started dating my (now) husband, and started to learn about farming and agriculture on his parent's farm. Before we were married I started my career selling print advertising and my husband had a job servicing dairy equipment. Marketing made sense to me and sales was something that I was a natural at, my husband had left the farm after high school and enjoyed working in agriculture but he hadn't always dreamt of returning to the farm. I have always had a deep love for animals and I grew up

C. Mess (✉)
Lake Mills, WI, USA
e-mail: carriemess@gmail.com

© Springer International Publishing AG 2017
L.S. Privalle (ed.), *Women in Sustainable Agriculture and Food Biotechnology*,
Women in Engineering and Science, DOI 10.1007/978-3-319-52201-2_9

digging in the dirt, planting vegetables and flowers with my grandmothers. My love for animals and seeing the seeds that I planted grow led to the big discussion a few months after our wedding about what was next for our lives. While many newlywed couples start dreaming of starting a family, instead my husband and I started to ask ourselves if we would be happy leaving our jobs in town and returning to the farm. Six months after we were married, we sat down at the kitchen table with his parents and we asked them to hire me.

At the time I really didn't know anything about cows or crops or farming in general. My Mother in law and Father in law were rightfully very nervous to hire someone with so little experience. Honestly, even I wasn't really sure that I could do the job. I've always been a night owl and the idea of having to get up every day at 5 am to milk cows was not something I was looking forward to. I knew how to drive a manual transmission but driving a tractor was very intimidating. While my in-laws didn't immediately jump at the chance to hire me, eventually I convinced them to give me a try. Thankfully for them, for me and most of all for our cows, I learned quickly and found my true passion on the farm. While my husband stayed working at his town job for several more years before eventually joining the farm as well, I worked alongside his parents learning from them and the different agriculture professionals that they worked with. Today it has been 8 years since I left my job in town and my husband and I are now in a formal partnership with his parents on our farm. My husband and I are the primary decision makers while we work through the transition of ownership of the farm as his parents get closer to retiring.

While many farms out there have had members of the same family work the land for generations, our farm is not one of them. My mother-in-law Cathy, grew up on a dairy farm thirty miles from where we farm now. My father-in-law Clem, grew up on a dairy farm 30 miles in the opposite direction. Both of their families sold their original farms for different reasons and purchased farms within one mile of the farm we have today in the few years after Cathy and Clem had both graduated high school. While their parents and siblings started farming land that was new to them, Clem and Cathy both branched out. Cathy went to college, majoring in architecture. Clem served in the Vietnam war and in 1972 after his return, he and one of his brothers purchased the farm we are on today. Being the two new families in the area, Clem and Cathy met each other and were married in 1973. The partnership between the brothers and their wives didn't work out and in 1981, the start of the hardest decade on farmers since the great depression, Clem and Cathy became the sole owners of the farm.

DronePhoto3: A 2014 aerial photo of our dairy farm and some of the land we grow crops on

Today when I stand in the dooryard of our farm, I see what most people would consider a typical dairy farm. In 2015, there was 9900 dairy farms in Wisconsin and the average farm had 129 dairy cows who made 22,697 lbs (2639 gallons) of milk per cow, annually.[1] On our farm we have around 110 milk cows, we raise around 120 heifers that will be our future dairy cows and we grow crops to feed our herd on about 300 acres of land. The numbers that quickly describe our farm are at the average, our herd of dairy cows has above average milk production, with our cows reaching close to 28,000 lbs of milk per cow, annually. While close management of our herd has helped us achieve higher levels of milk production, the biggest reason for our cow's above average productivity is because our farm has always been very progressive in trying new ideas and adopting new technology that benefits our cows and our land.

With 300 acres of land, we are able to grow almost all of the feed our cows and heifers eat. We do purchase and bring in cottonseed, which we use as a source of energy for our cows and brewer's grains; a byproduct of beer making that is also great feed for our cows. In addition to those two ingredients our cows eat a mixture of the crops we grow. The three main crops we grow are corn, both for silage, which is made up of the whole corn plant and for shell corn, which is just the kernel, we also grow alfalfa and soybeans. In addition to those main crops, depending on the year and what type of feed we need for our cows, we grow wheat, rye, peas, oats, triticale (a wheat and rye hybrid) and have even experimented with sunflowers and a sorghum sudangrass hybrid. For the

---

[1]Source: http://media.eatwisconsincheese.com/assets/images/pdf/WisconsinDairyData.pdf

past 4 years we have also held back an acre of our land from regular crop production and have used that land to grow sweet corn that is donated to our local Second Harvest Food Bank.

CornDonation: A photo of me dropping off a tote of sweet corn at Second Harvest Food Bank that we grew in order to donate

Each of the crops we grow has a different nutritional makeup and fits into our cow's diet in different ways. We work with a ruminant nutritionist to analyze each crop that we grow and to formulate the perfect combination of our available feed for our cows. What our cows eat each day is a precise blend of the crops we so carefully grew and harvested. Each day our milk cows eat over 100 lbs of feed per cow, containing corn silage, alfalfa, brewer's grains, cottonseed, high moisture shell corn, soybean meal and a vitamin and mineral mix. Every batch of feed mixed is measured to the pound. The milk cows have a different recipe and ingredients than our dry cows who are pregnant and not being milked. The older heifers have a different mix of ingredients than our younger heifers, our youngest heifers have a completely different mix. Each animal on our farm is fed exactly what she needs for the stage of growth or production she is in. The precision we have with making the best feed for our cows doesn't start when we mix a batch, it starts when we plant the seeds to grow the crops.

Our farm is situated on a fairly flat tract of land that runs along the Crawfish river. The soil is dark and rich and the clay is deep under the good topsoil, in general these qualities make our area prime land for growing crops like corn, soy and alfalfa. Our farm's proximity to the river means that we have more of a mixed bag when it comes to growing conditions. In wet years, the springs in some of our fields keep us from planting large areas, or if the rain comes after the planting, it can drown out acres of plants. In dry years, some of our marginal land will have the best yields. On an average year, the 300 acres of crops we grow will provide enough feed for our cows plus a little extra to sell or store. On a wet year or a drought year, we may not be able to grow enough to crops and would have to purchase additional feed. Without much cushion in our acreage, It's vital for us to manage our land carefully, to ensure that we have enough feed for our cows to last from one harvest to the next.

Drone Photo1

Drone Photo2: These drone photos are of our land that runs along the river. This photo was taken in 2014, a wetter than normal year. You can see the large areas of our corn crop that was drowned out after planting by flooding. You can also see the corn that survived is stressed and not growing as well

To truly understand the role that GMO technology has played on our farm, we have to look back at the decisions that led us to choose to plant modified seeds. The first GMO seeds we ever planted on our farm were RoundUp Ready Soybeans. Introduced to the market in 1996, the soybeans were modified in a way that allowed them to be sprayed with glyphosate and not die. This allowed for better weed control while reducing the use of other harsher chemicals and reducing the need for tillage in fields. When the RoundUp Ready soybeans became available, trials showed that they actually didn't produce as well as non GMO hybrids. Despite the initial performance concerns, our crop consultant suggested that Clem and Cathy try them out in some of our more difficult land closest to the river. The idea of trying the most expensive seeds our farm had ever purchased on some of our worst land for growing crops, knowing that we might get less of a crop was risky. However, being able to control weeds without tilling the soil that was often so wet, our tractors would get stuck and getting away from using herbicides with higher risk factors so close to the river was enough of a benefit that the decision to try the new technology out was worth the risk.

With the focus of the best plant genetics shifted to the GMO soybean families, in the 20 years since GMO soybeans were introduced to the market, the average yields from the GMO soybeans have caught up to the yields of non GMO beans. After planting those first GMO seeds 20 years ago, Clem and Cathy were convinced that despite the initial seed cost, the improved weed control available with a single application of glyphosate versus prior years that required more tillage and more herbicide applications, the GMO soybeans were a much better option for our land.

A few years after glyphosate resistant soybeans came to our farm, we planted our first GMO corn. Today GMO seed corn is available with not only herbicide resistance built in, it is also available with resistance to both below ground pest and above ground pests such as the Corn Rootworm and Corn Borers. While the built in herbicide resistance is helpful for all the same reasons GMO soybeans worked well for us, the ability to control common pests was a new advantage that we had to evaluate on our farm.

While soybeans are a portion of our cow's diets, corn is a much more significant part of what our cows eat each day. While many people know that cows eat corn, most people don't realize the entire corn plant used is for feed. The corn plant is chopped into small pieces and stored in silos, bunkers or bags where it ferments and is fed throughout year. With corn silage as a major component of our cow's diet, it is even more important for us to protect the whole plant from pests.

It's important to understand the life cycle of a corn plant to understand the challenges that pests create for farmers. A corn plant has seven major parts. The roots, brace roots, stalk, leaf, silk, tassel and ear. During pollination the tassel produces pollen which is caught by the silk at the end of the ear. Each individual silk is attached to the ear where a single corn kernel will develop after pollination. A corn plant's growth is classified in stages from VE, V1–V10 and VT. When the corn plant first emerges from ground (stage VE) it sends up a single spike that splits into separate leaves (stage V1). The plant continues to add leaves; however the growing point of the plant is still underground. Once the plant's growing point is above ground (stage V6) the plant is well established. Ear shoots and the tassel begin to develop (stage V10) in preparation for pollination. Once the tassel appears (stage VE) all the leaves the plant will grow are present and the plant's energy is focused on reproduction. After successful pollination occurs and the plant creates its seed, which we know as corn kernels, the plant dies and all parts start to dry. Farmers harvest the corn after the plant has completed its life cycle.

When corn first comes out of the ground, the main pest we have to deal with is weeds. We can either manage weeds with an herbicide application, by using a tractor and cultivator to turn the dirt between the corn rows or as some organic farms do, use flame to burn back weeds. Once the corn grows tall enough it creates a canopy that blocks most sunlight and keeps weeds from flourishing.

Cornandweeds: A corn field, prespraying that is being taken over by weeds

CultivatingCorn: To control weeds by cultivating, a farmer must drive carefully between corn rows with the tractor and cultivator. The cultivator kills weeds by turning the dirt between the rows of corn

FlameWeedingCorn: Flame weeding relies on a propane flame that burns the weeds growing between corn rows. The corn is also burned but since the flaming happens early in its development, the corn plant is able to recover

Corn rootworm and corn borers are the two main pests in our area that affect corn yields. Both feed on corn plants at different stages of their development. Corn rootworm is a small beetle and in the larval stage, the worm feeds on the plants' roots, which keeps the plant from getting enough water and nutrients to grow well. Later in the adult stage the beetle feeds on the plant leaf and the silk. Without the silk, the corn cannot pollinate and no kernels will form on the ear.

Corn borers cause a different kind of damage. Over the cycle of a single growing season, corn borers will have multiple generations. As a moth in the adult stage, the corn borer doesn't do much damage to the corn plant. However the adult moth lays eggs on the growing corn and when the eggs hatch, the problems start. Each generation hatched in a growing season of corn borer larvae attacks different parts of the plant. The early generation will eat leaves and burrow into the rib of the leaf, causing it to break. The second generation larvae hatch and move to the point of the plant where the leaf meets the stalk and eats pollen and plant tissue. Both generations will travel to the stalk and bore into it. The larvae pupate inside the corn stalk and emerge as adults later in the summer. The action of the larvae eating holes and boring into the stalk not only weakens the stalk it also disrupts the plant's delivery system for nutrients and water. As the corn plant dries down after it is done growing, a plant infested with corn borers will have a weakened stalk that will break and fall, this is called stalk lodging. When a corn plant lodges, harvesting equipment is unable to reach the full plant and a reduced harvest is the result.

With the insect challenges happening throughout the growing season, pest control quickly becomes tricky. With our corn being 10'–13' tall when it tassels, sprayers aren't tall enough to travel through our fields. Spraying from crop dusting planes or helicopters can be done but with the thick canopy of corn leaves and insects that bore into the plant, the pesticide doesn't always make it to where the bugs are. Leaving few effective options to decrease insect pressure on our corn crop.

When corn that was engineered to be glyphosate resistant and also contained Bacillus thuringiensis, a soil bacterium that is deadly to the corn borer and corn rootworm came onto the market, it was an easy decision for Clem and Cathy to make to switch from the conventional seed they had always grown to the new GMO seeds being offered. Through the years of growing GMO crops we have seen a drastic decrease in the amount of pesticides we use as well as the amount of tillage we need to control weeds.

From the business perspective, the decrease in the amount of spraying we had to do to control weeds and pests in our soybean and corn crop and the increased yields in our corn harvest helped our bottom line. Eliminating the need for extensive tillage for weed control pre planting and cultivating corn rows after the corn was up but not yet tall enough to block out weeds, saved us hours of work each season as well as diesel fuel and wear and tear on our tractors. Looking beyond our farm's bottom line and at the greater picture, our change from traditional seed choices to GMO technology has meant less chemicals used and has reduced our farm's carbon footprint, benefiting all of us.

PlantingCorn: Our employee Jolene planting corn for the first time in one of our fields

The decision to switch from conventionally bred soybean and corn seed to their genetically modified counterparts happened long before I ever stepped foot on the farm. However, I have been able to see the difference in how we manage our alfalfa crop between fields that are planted with genetically modified seed and those that are not, first hand.

RoundUp Ready alfalfa was first introduced in 2005, but wasn't widely available until 2006. We first planted RoundUp Ready Alfalfa that year. In 2007 Monsanto was forced to pull RoundUp Ready alfalfa from the market after the USDA was sued by the anti GMO group, the Center for Food Safety. The suit filed with the US District Court of Northern California alleged that the USDA did not prepare an Environmental Impact Statement (EIS) for the Roundup Ready Alfalfa and that went against the USDA procedures. Farms like ours that had already planted the seed were not forced to destroy our crop, like many thought might happen, however we did have to register GPS coordinates for each of our fields that had been planted with Roundup Ready Alfalfa.

The District Court's decision was appealed and in 2010 the US Supreme Court overturned the decision of the lower court, allowing farmers to plant Roundup Ready alfalfa while the Environmental Impact Statement was being prepared rather than having to wait until the EIS was completed. Several follow up lawsuits were filed to continue blocking the technology but none succeeded in doing so. In 2011

Roundup Ready alfalfa was reintroduced to the market however the availability of seed was an issue for several years.

To really understand the change that glyphosate resistant alfalfa made for our farm, you should have some details about how we grow and harvest alfalfa. Unlike corn or soybeans, an annual crop that is planted in spring and harvested in the fall, alfalfa is a perennial that is planted in a stand that we will harvest 3 or 4 cuttings off of each season for 3–4 years. While a field of alfalfa can continue on for much longer than the 3 years our fields average, in the upper midwest and other areas with harsh winter conditions, it's common to have some plants killed off each winter and as the plant ages, growth isn't as vigorous and yields decline. The 1st year an alfalfa field is planted, we will only take two cuttings, allowing the plant to become well established. The 2nd year of an alfalfa stand is usually the best for yields and the 3rd year quality and yields drops off as more plants die and allow weeds to grow in their place.

We planted our first Roundup Ready Alfalfa seed after it's reintroduction to the market last spring. I remember my husband and I having a conversation about the recommendation from our crop consultant to plant two of our fields with Roundup Ready alfalfa. I wasn't exactly sold on the idea of spending significantly more on GMO alfalfa seed, rather than the standard seed we had planned on planting and my husband wasn't sure it was the best option either. We called our crop consultant while on speaker phone to discuss his reasoning on making that recommendation. He explained that the two fields we were looking at planting in alfalfa were both under heavier than normal weed pressure and he thought it best to use seed that would help us control the weeds. After talking it over, we decided to plant one field with the glyphosate resistant variety and the other with conventional alfalfa seed.

Before planting both fields were tilled and an herbicide was applied to kill any weeds. To help the conventional alfalfa get off to the best start we planted it with a cover crop of peas and oats. This is a standard practice on fields of non GMO alfalfa. The cover crop grows quickly choking out any weeds while underneath the alfalfa can get established. Since the cover crop plants aren't perennial, once the field has been harvested once, the cover crop is gone and the alfalfa is established enough to better handle weed pressure. While a cover crop not only provides additional feed for our cows and helps prevent weed growth, they can cause problems as well.

When cutting alfalfa for harvest, timing is everything to get the best quality. Along with corn silage, alfalfa is the bulk of what our cows eat each day. The higher the quality of alfalfa we harvest, the better feed it makes for our cows and better feed makes more milk. Last summer when it came time for us to cut the two fields of new seeding of alfalfa, mother nature decided not to cooperate with us. In order to harvest alfalfa we need a few days of dry weather and instead we just kept getting rain. For the field of Roundup Ready alfalfa the wet weather meant that while that field was going to produce lower quality feed by the time we were able to harvest it, the plants themselves would be just fine. However our field with the cover crop was in trouble.

The rain delaying harvest on our cover crop allowed the crop to continue growing and soon the peas and oats were choking out the young alfalfa growing under it. As the rainy weather continued, the cover crop which was tall and lush, got heavy and started to create dense mats of plants, further choking the alfalfa. By the time we were finally able to harvest the field, many of the alfalfa plants were dead. Not only did this decrease our harvest last year, it also allowed weeds to take over the areas where the alfalfa had died, which reduces the quality of our crop, and this field will have reduced yields and quality for its life span.

PoorAlfalfa: This photo shows two alfalfa fields that have real problems with weeds. In the top photo, you can see all the dandelions that have gone to seed in the field. In the bottom photo you can see where our cover crops killed off the alfalfa and allowed weeds to take over. Both fields will have reduced yields and greatly reduced quality because of the weeds

Shortly after joining my in-law's on the farm, I joined the movement of farmers that were using social media to connect with the public who wanted to learn more about where their food came from. As a freshly minted "Agvocate", as I learned about farming myself, I shared my new knowledge on twitter, facebook and my blog. Before joining the farm, I knew about GMO's but my knowledge was very

limited. I wasn't really for or against them, I doubted that my in-law's would plant
something that wasn't useful. However, seeing a lot of chatter online from people
who were fighting against GMO technology made me wonder if their wasn't more
to be concerned about than their overall usefulness to farmers.

When I joined the farm the only way I could handle learning all at once, what
most dairy farmers had spent their entire lives learning was to focus on one aspect at
a time. So I first focused on our cows and one at a time I focused on learning as
much as I could about vaccination protocols, udder health, reproduction, lameness,
calf health and a mountain of other aspects of caring for cows. Learning about our
cows was easy, my love for our animals created a driving force to understand what
made cows tick helped me to easily understand the science of cows. Learning in a
hands on way worked well for me and I had access to our veterinarians, nutritionist
and other experts to answer my questions. But when I decided to learn about GMO
technology in plants, the same feelings of anxiety and confusion that I had in school
started to crop back up. Genes weren't something I could see, or touch. The science
was overwhelming to me and even more overwhelming were the online arguments.
So once again I decided to break it down into more manageable parts.

Before I got to learning about the technology itself, I decided to look at some of
the claims I had seen online about GMO's that weren't necessarily specific to the
technology. From lawsuits filed against farmers for saving seeds or for pollen
drifting into their fields, to farmers in India committing suicide because of GMO
crops to GMO wheat causing the rise in gluten intolerance, I started by reading as
much as I could on each topic from both sides of the debate. I talked to other
farmers about saving seeds and the technology agreements they signed when they
purchased patented seeds. As I read and listened, I learned that the amount of false
information on the topic of GMO crops was pretty mind boggling, and a great deal
of the misinformation spreading like wildfire across the internet came from the
average person not understanding how food is grown or produced on farms. At the
same time I also saw farmers who grow GMO crops, throwing Organic farmers
who didn't use GMO technology under the bus. I could see how the average person
was easily confused on the topic and why many people were so quick to jump on
the bandwagon against GMOs.

Once I had a better understanding on the hot button issues, I started to focus on
the science side of how GMOs are created and what the difference was between a
GMO and a hybrid. I started looking at what GMO technology does beyond making
crops that are herbicide and pest resistant. At first I went to science minded websites
and when I struggled to understand something, I called my best friend who ran a lab
that worked with genes and proteins. As I gained confidence in my base knowledge
I dug further and started reading reports and studies, again from both sides of the
issue. When I decided to learn more about glyphosate, not only did I read studies, I
talked to the man that we hire to spray our fields. His 30+ years in business, gave
me insight into the difference in chemical application rates over the years. He told
me about the different chemicals he uses not only becoming more efficient over the
years and safer to handle, but how technology in spraying equipment has advanced
and helped use even less chemicals. I talked to our crop consultant about the

challenge of herbicide resistant weeds and the effect of pesticides on honey bees and monarch butterflies. While I still strive to learn more, after looking at the issue and the science, I believe that GMO technology is a safe and positive addition to our farm.

The future of our farm is something my husband and I talk about regularly. After having our first child last summer, our focus towards the future is renewed with the hope that one day, our children will farm the same land that their parents and grandparents did. With an eye towards the future and the economic realities of dairy farming, we know that at some point we will most likely choose to grow our herd of cows. With crop land in our area being expensive and lots of competition for it, we know that we will need to do what we can to increase the amount of feed we can grow on the land we have available to us. I believe that GMO technology will be a significant part of what helps us to achieve a higher level of productivity from our land.

This year a new variety of alfalfa, genetically modified to have lower levels of lignin has been introduced. Lignin is found in all vascular plants and is what gives fruits and veggies their crunch and fiber. Like biting into an overripe and woody radish, as alfalfa matures, lignin increases and causes the alfalfa to become too fibrous and less digestible for our cows. By lowering the lignin levels in alfalfa, we can allow our alfalfa to grow for longer, which increases the amount of feed we can make per acre, without decreasing the quality of our crop. While we have not planted this seed ourselves, I can certainly see where this technology would help us grow more feed on our current land.

Somewhere in a lab, the women in this book and many others are working on genetically modified varieties of corn that can handle wet soil or drought conditions. When these seeds hit the market, our farm will once again have the potential to increase our land's yields in our soggy river bottom land. Likewise, the drought resistant varieties should help on our highest fields where retreating glaciers left ribbons of sandy soil that doesn't hold moisture. Work in labs to figure out how to keep alfalfa from mildewing or being eaten by leafhoppers will further reduce our farm's use of chemicals.

Beyond our farm dooryard, GMO technology is helping farmers of crops beyond the soybeans, corn and alfalfa we grow achieve the same goals we have of doing more with less while reducing the need for chemicals. The same Bt technology that protects our corn from insects when used in cotton, allows cotton farmers to reduce their use of pesticides. Canola and sugar beet farmers are now able to control weeds in their crops with glyphosate resistant varieties. Genetic modification to crops goes beyond weed and insect control. Genetic modifications to the papaya to increase it's resistance to ringspot virus, saved the Hawaiian papaya industry. A recently approved apple variety resists browning helping to reduce food waste, while newly introduced types of potato resists browning and produces less of the carcinogen acrylamide when it's fried. The technology and science involved in genetic modification is rapidly changing how farmers of all types produce their crops, I believe for the better.

Each summer in my garden I plant carefully selected varieties of heirloom tomatoes. I select plants based on their past performance, their color, their size and most of all their taste. When we select seed for our crops we consider past performance, the qualities of the field we are planting, the piece that specific crop will add to our cow's diet and how we will control the weeds and pests in the crop. Just as I appreciate the women who saved the seeds for my heirloom tomatoes from their gardens generation after generation, I appreciate the women who have done the work to create the choices we have today in our crops and the choices we will have tomorrow. Their work is my work.

Index

A
Aaron deGrassi, 126
Africa, 40, 51
Africa Biofortified Sorghum (ABS) Project,
    The, 131
African Union (AU), 129
Agrichemical, 71, 76, 80
Agricultural research, 129
Agriculture, 67, 68, 88, 90, 103, 111, 112
Agriculture toxic, 74
Agrobacterium, 21, 24–27, 31–33
Ann Depicker, 12
Anne Knupp Crossway, 11
Anti-GM, 125–128

B
Bacillus thuringiensis (Bt), 8, 9, 14, 51, 59, 73,
    74, 78, 77, 90, 128, 144
    Bt crop(s), 61, 62, 68
    Bt plants, 61, 62
    Bt proteins, 59–61, 63, 64
    Bt resistance, 61
Bacteria, 10, 25, 26, 30, 32
Bacterium, 86
Barbara Hohn, 8, 9
Barbara Mazur, 11
Barbara McClintock, 5, 9, 10
Biochemistry, 59, 63, 71, 865–88, 92, 94
Biology, 85–89, 92, 94
    microbiology, 11
    plant biology, 1, 11, 12
Biopharmaceutical(s), 115, 116, 119
Bioprocessing, 108, 115, 118, 122
Biotech, 71, 74, 75, 77, 79–81, 128, 131
Biotechnology, 1, 7, 8, 11–14, 39, 48–53, 55,
    59, 61, 71, 76, 80, 82, 85, 87, 89, 90, 93,
    94, 97, 98, 101–105, 108, 109, 111, 114,
    120

Biowatch, 128
BTEC, 118, 122
Btk, 60, 61, 63
Bt protein, 73–76, 78
Burkina Faso, 129

C
Canada, 75, 77
Center, The, 97–103, 105, 107, 109, 118, 120,
    122
Climate, 36
Comprehensive African Agricultural
    Development Program (CAADP), 129
Corn, 73, 75, 76, 78, 137–139, 141, 144, 146,
    149
Croplife International (CLI), 129
Crops, 136–140, 144, 148–150
Cry9C, 65, 66

D
DNA, 6–11, 24–29, 31–33, 39, 40, 42–46, 75
DNA. Genes, 41, 42
DNA r-DNA, 47
Dr Ben Odhiambo, 126
Dr Francis Nang'ayo, 126
Drought, 139, 149

E
Education, 98, 100, 103, 104, 106, 108, 111,
    114, 115, 117, 119, 120, 122
Education grant, 120
Education project(s), 97, 103, 105
Egypt, 129, 131
Elizabeth Hood, 12
Enzyme, 24, 25, 28, 29, 31, 61, 63
EPA, 48, 49, 51, 75–78, 90, 91
Estella Eleanor Carothers, 4
Escherichia coli, 88

© Springer International Publishing AG 2017
L.S. Privalle (ed.), *Women in Sustainable Agriculture and Food Biotechnology*,
Women in Engineering and Science, DOI 10.1007/978-3-319-52201-2